Marx Joyce
Abbott Hardy Machiavelli Chesterton Emerson Austen
Defoe Melville Montaigne Cooper Hugo
Haggard Eliot
Carroll Christie Grimm
Stoker Molière
Wilde Maupassant Byron Schiller
Engels
Garnett Fitzgerald Smith Kafka
Goethe Hawthorne
Cotton Einstein Dostoyevsky Hall
Baum Kipling Doyle Willis
Henry
Leslie Dumas Nietzsche
Flaubert Turgenev Balzac
Stockton Vatsyayana Crane
Burroughs Verne
Curtis Tocqueville Gogol Vinci
Homer Widger Tolstoy Whitman Busch
Darwin Thoreau
Potter Freud Zola Twain Scott
Kant Jowett Lawrence Plato Harte
Stevenson Dickens Hesse
Andersen Burton
London Descartes Cervantes
Poe Aristotle Wells Voltaire
Hale James Hastings Cooke
Bunner Shakespeare Irving
Richter Chambers
Doré da Benedict Alcott
Dante Shaw Pushkin
Swift Chekhov Newton
Wodehouse

tredition®

tredition was established in 2006 by Sandra Latusseck and Soenke Schulz. Based in Hamburg, Germany, tredition offers publishing solutions to authors and publishing houses, combined with worldwide distribution of printed and digital book content. tredition is uniquely positioned to enable authors and publishing houses to create books on their own terms and without conventional manufacturing risks.

For more information please visit: www.tredition.com

TREDITION CLASSICS

This book is part of the TREDITION CLASSICS series. The creators of this series are united by passion for literature and driven by the intention of making all public domain books available in printed format again - worldwide. Most TREDITION CLASSICS titles have been out of print and off the bookstore shelves for decades. At tredition we believe that a great book never goes out of style and that its value is eternal. Several mostly non-profit literature projects provide content to tredition. To support their good work, tredition donates a portion of the proceeds from each sold copy. As a reader of a TREDITION CLASSICS book, you support our mission to save many of the amazing works of world literature from oblivion. See all available books at www.tredition.com.

 Project Gutenberg

The content for this book has been graciously provided by Project Gutenberg. Project Gutenberg is a non-profit organization founded by Michael Hart in 1971 at the University of Illinois. The mission of Project Gutenberg is simple: To encourage the creation and distribution of eBooks. Project Gutenberg is the first and largest collection of public domain eBooks.

The Naturalist on the Thames

C. J. (Charles John) Cornish

Imprint

This book is part of TREDITION CLASSICS

Author: C. J. (Charles John) Cornish
Cover design: Buchgut, Berlin – Germany

Publisher: tredition GmbH, Hamburg - Germany
ISBN: 978-3-8424-6574-9

www.tredition.com
www.tredition.de

Copyright:
The content of this book is sourced from the public domain.

The intention of the TREDITION CLASSICS series is to make world literature in the public domain available in printed format. Literary enthusiasts and organizations, such as Project Gutenberg, worldwide have scanned and digitally edited the original texts. tredition has subsequently formatted and redesigned the content into a modern reading layout. Therefore, we cannot guarantee the exact reproduction of the original format of a particular historic edition. Please also note that no modifications have been made to the spelling, therefore it may differ from the orthography used today.

PREFACE

Having spent the greater part of my outdoor life in the Thames Valley, in the enjoyment of the varied interests of its natural history and sport, I have for many years hoped to publish the observations contained in the following chapters. They have been written at different intervals of time, but always with a view to publication in the form of a commentary on the natural history and character of the valley as a whole, from the upper waters to the mouth. For permission to use those which have been previously printed I have to thank the editors and proprietors of the *Spectator*, *Country Life*, and the *Badminton Magazine*.

C.J. CORNISH.

ORFORD HOUSE, CHISWICK MALL.

CONTENTS

THE THAMES AT SINODUN HILL

THE FILLING OF THE THAMES

THE SHELLS OF THE THAMES

THE ANTIQUITY OF RIVER PLANTS

INSECTS OF THE THAMES

"THE CHAVENDER OR CHUB"

THE WORLD'S FIRST BUTTERFLIES

BUTTERFLY SLEEP

CRAYFISH AND TROUT

FOUNTAINS AND SPRINGS

BIRD MIGRATION DOWN THE THAMES

WITTENHAM WOOD

SPORT AT WITTENHAM
   SPORT AT WITTENHAM (*continued*)

A FEBRUARY FOX HUNT

EWELME — A HISTORICAL RELIC

EEL-TRAPS

SHEEP, PLAIN AND COLOURED

SOME RESULTS OF WILD-BIRD PROTECTION

OSIERS AND WATER-CRESS

FOG AND DEW PONDS

POISONOUS PLANTS

ANCIENT THAMES MILLS

THE BIRDS THAT STAY

ANCIENT HEDGES

THE ENGLISH MOCKING BIRD

FLOWERS OF THE GRASS FIELDS

RIVERSIDE GARDENING

COTTAGES AND CAMPING OUT

NETTING STAGS IN RICHMOND PARK

RICHMOND OLD DEER PARK

FISH IN THE LONDON RIVER

CHISWICK EYOT

CHISWICK FISHERMEN

BIRDS ON THAMES RESERVOIRS

THE CARRION CROW

LONDON'S BURIED ELEPHANTS

SWANS, BLACK AND WHITE

CANVEY ISLAND

THE LONDON THAMES AS A WATERWAY

THE THAMES AS A NATIONAL TRUST

LIST OF ILLUSTRATIONS

A FOX FLUSHING PHEASANTS

WILD DUCK

A FULL THAMES

SHELLS OF THE THAMES

A FLOWERY BANK

BURR REED AND FLOWERING RUSH

A MONSTER CHUB

BUTTERFLIES AT REST

A TROUT

OTTERS

A WATERHEN ON HER NEST

A DABCHICK

A BADGER

FOX AND CUB

EWELME POOL

A NIGHTJAR AND YOUNG ONE

A REED-BUNTING

PEELING OSIERS

BOTLEY MILL

EEL BUCKS

ORCHIS

WATER VIOLET AND WILD IRIS

A NETTED STAG

BREAM AND ROACH

A GRAMPUS AT CHISWICK

SMELTS

THE LOBSTER SMACK INN, CANVEY ISLAND

THE STEPPING-STONES AT BENFLEET

HAULING THE NETS FOR WHITEBAIT

FISHING BOATS AT LEIGH

THE NATURALIST ON THE THAMES

THE THAMES AT SINODUN HILL

Fresh water is almost the oldest thing on earth. While the rocks have been melted, the sea growing salter, and the birds and beasts perfecting themselves or degenerating, the fresh water has been always the same, without change or shadow of turning. So we find in it creatures which are inconceivably old, still living, which, if they did not belong to other worlds than ours, date from a time when the world was other than it is now; and the fresh-water plants, equally prehistoric, on which these creatures feed. Protected by this constant element the geographical range of these animals and plants is as remarkable as their high antiquity. There are in lake Tanganyika or the rivers of Japan exactly the same kinds of shells as in the Thames, and the sedges and reeds of the Isis are found from Cricklade to Kamschatka and beyond Bering Sea to the upper waters of the Mackenzie and the Mississippi. The Thames, our longest freshwater river, and its containing valley form the largest natural feature in this country. They are an organic whole, in which the river and its tributaries support a vast and separate life of animals and plants, and modify that of the hills and valleys by their course. Civil law has recognised the Thames system as a separate area, and given to it a special government, that of the Conservators, whose control now extends from the Nore to the remotest springs in the hamlets in its watershed; and natural law did so long before, when the valley became one of the migration routes of certain southward-flying birds. Its course is of such remote antiquity that there are those who hold that its bed may twice have been sunk beneath the sea, and twice risen again above the face of the waters.[1] It has ever been a masterful stream holding its own against the inner forces of the earth; for where the chalk hills rose, silently, invisibly, in the long line from the vale of White Horse to the Chilterns the river seems to have worn them down as they rose at the crossing point at Pangbourne, and kept them under, so that there was no barring of the Thames, and no subsequent splitting of the barrier with gorges, cliffs, and falls. Its clear waters pass from the oolite of the Cotswolds, by the blue lias and its fossils, the sandstone rock at Clifton Hampden, the gravels of Wittenham, the great chalk range of the downs, the greensand, the Reading Beds, to the geological pie of the London Basin, and the beds of drifts and brick earth in which lie bedded the frames and fragments of its prehistoric beasts. In and beside its valley are great woods, parks, downs, springs, ancient mills and fortresses, palaces and villages, and such homes of prehistoric man as Sinodun Hill and the hut remains at Northfield. It has 151 miles of fresh water and 77 of tideway, and is almost the only river in England in which there are islands, the famous eyots, the lowest and largest of which at Chiswick touches the London boundary.

After leaving Oxford the writer has lived for many years opposite this typical and almost unspoilt reach of the London river, and for a considerable time shot over the estate on the upper Thames of which Sinodun Hill is the hub and centre. This fine outlier of the chalk, with its twin mount Harp Hill, dominates not only the whole of the Thames valley at its feet, but the two cross vales of the Thame and the Ock. On the bank opposite the Thame joins the Isis, and from thence flows on the THAMES. Weeks and months spent there at all seasons of the year gave even better opportunities for becoming acquainted with the life of the Upper Thames, than the London river did of learning what the tidal stream really is and may become. Fish, fowl and foxes, rare Thames flowers and shy Thames chub, butterflies, eel-traps, fountains and springs, river shells and water insects, are all parts of the "natural commodities" of the district. There is no better and more representative part of the river than this. Close by is Nuneham, one of the finest of Thames-side parks, and behind that the remains of wild Oxfordshire show in Thame Lane and Clifton Heath. How many centuries look down from the stronghold on Sinodun Hill, reckoning centuries by human occupation, no one knows or will know. There stands the fortress of some forgotten race, and below it the double rampart of a Roman camp, running from Thame to Isis. Beyond is Dorchester, the abbey of the oldest see in Wessex, and the Abbey Mill. The feet of the hills are clothed by Wittenham Wood, and above the wood stretches the weir, and round to the west, on another great loop of the river, is Long Wittenham and its lovely backwater. Even in winter, when the snow is falling like bags of flour, and the river is chinking with ice, there is plenty to see and learn, or in the floods, when the water roars through the lifted hatches and the rush of the river throbs across the misty flats, and the weeds and sedges smell rank as the stream stews them in its mash-tub in the pool below the weir.

[1] Phillips, "Geology of Oxford and of the Valley of the Thames."

THE FILLING OF THE THAMES

In the late autumn of 1893, one of the driest years ever known, I went to the weir pool above the wood, and found the shepherd fishing. The river was lower than had ever been known or seen, and on the hills round the "dowsers" had been called in with their divining rods to find the vanished waters.

"Thee've got no water in 'ee, and if 'ee don't fill'ee avore New Year, 'ee'll be no more good for a stree-um"! Thus briefly, to Father Thames, the shepherd of Sinodun Hill. He had pitched his float into the pool below the weir—the pool which lies in the broad, flat fields, with scarce a house in sight but the lockman's cottage—and for the first time on a Saturday's fishing he saw his bait go clear to the bottom instead of being lost to view instantly in the boiling water of the weir-pool. He could even see the broken piles and masses of concrete which the river in its days of strength had torn up and scattered on the bottom, and among them the shoals of fat river fish eyeing his worm as critically as his master would a sample of most inferior oats. Yet the pool was beautiful to look upon. Where the water had sunk the rushes had grown taller than ever, and covered the little sandbanks left by the ebbing river with a forest of green and of red gold, where the frost had laid its finger on them. In the back eddies and shallows the dying lily leaves covered the surface with scales of red and copper, and all along the banks teazles and frogbits, and brown and green reeds, and sedges of bronze and russet, made a screen, through which the black and white moorhens popped in and out, while the water-rats, now almost losing the aquatic habit, and becoming pedestrian, sat peeling rushes with their teeth, and eyeing the shepherd on the weir. Even the birds seemed to have voted that the river was never going to fill again, for a colony of sandpipers, instead of continuing their migration to the coast, had taken up their quarters on the little spits of mud and shingle now fringing the weir-pool, and were flitting from point to point, and making believe it was a bit of Pagham Harbour or Porchester Creek. On every sunny morning monster spiders ran out from the holes and angles of the weir-frame, and spun webs across and across the straddling iron legs below the footbridge, right down to the lowered surface of the water, which had so sunk that each spider had at least four feet more of web than he could have reckoned upon before and waxed fat on the produce of the added superficies of enmeshed and immolated flies. So things went on almost till New Year's Eve. The flats of the Upper Thames, where the floods get out up the ditches and tributaries, and the wild duck gather on the shallow "splashes" and are stalked with the stalking-horse as of old, were as dry as Richmond Park, and sounded hollow to the foot, instead of wheezing like a sponge. The herons could not find a meal on a hundred acres of meadow, which even a frog found too dry for him, and the little brooks and land-springs which came down through them to the big river were as low as in June, as clear as a Hampshire chalk stream, and as full of the submerged life of plants. Instead of dying with the dying year at the inrush of cold water brought by autumn rains, all the cresses, and tresses, and stars, and tangles, and laced sprays of the miniature growth of the springs and running brooks were as bright as malachite, though embedded in a double line of dead white shivering sedge. And thus the shortest day went by, and still the fields lay dry, and the river shrank, and the fish were off the feed; and though murky vapours hung over the river and the flats and shut out the sun, the long-expected rains fell not until the last week's end of the year. Then at last signs and tokens began by which the knowing ones prophesied that there was something the matter with the weather. The sheep fed as if they were not to have another bite for a week, and bleated without ceasing, strange birds flew across the sky in hurrying flocks, and in all the country houses and farmers' halls the old-fashioned barometers, with their dials almost as big as our eight-day clocks and pointers as long as a knitting-needle, began to fall, or rather to go backwards, further than was ever recorded. And whereas it is, and always has been, a fact well known to the owners of these barometers that if they are tapped violently in the centre of their mahogany stomachs the needle will jerk a little in the direction of recovery, and is thereby believed to exercise a controlling influence in the direction of better weather, the more the barometers were tapped and thumped the more the needle edged backwards, till in some cases it went down till it pointed to the ivory star at the very bottom of the dial, and then struck work and stuck there.

[Illustration: WILD DUCK. *From a photograph by Charles Reid.*]

[Illustration: A FULL THAMES. *From a photograph by Taunt & Co.*]

That night the storm began. To connoisseurs in weather in the meteorological sense it was a joy and an ensample, for it was a perfect cyclonic storm, exactly the right shape, with all its little dotted lines of "isobars" running in ovals one inside another. From another point of view it was the storm of an hour spread over two days, so that there was plenty of time to see and remember the normal ways of cyclones, which may be briefly described as first a flush of heat whether in summer or winter, then a furious wind, then hurrying clouds and much rain, with changes of wind, then more clouds and more rain, then a "clearing shower" with most rain, then a furling and brailing-up of the rain clouds, splashes of blue in the sky, with nets of scud crossing them, sudden gleams of sun, sudden cold, and perhaps a hail shower, and then piercing cold and sunlight. All which things happened, but took a long time about it. The storm began in the night, and howled through the dark. The rain came with the morning; but it was the "clearing shower," which lasted ten hours, which caused the filling of the Thames. The wind still blew in furious gusts, but the rain was almost too heavy to be moved. The sky was one dark, sombre cloud, and from this the rain poured in slanting lines like pencils of water. But across this blanket of cloud came darker, lower, and wetter clouds, even more surcharged with water, from which the deluge poured till the earth was white like glass with the spraying drops. Out in the fields it was impossible to see through the rain; but as the end of the column of cloud began to break and widen the water could be seen in the act of passing from the land to the river. On the fallows and under the fences all the surface earth was beaten down or swept away. All seeds which had sunk naturally below the surface were laid bare. Hundreds of sprouting horse chestnuts, of sprouting acorns beneath the trees, thousands of grains of fallen wheat and barley, of beans, and other seeds of the farm were uncovered as if by a spade.

Down every furrow, drain, watercourse, ditch, runnel, and watercut, the turbid waters were hurrying, all with one common flow, all with increasing speed, to the Thames. The sound of waters filled the air, dropping, poppling, splashing, trickling, dripping from leaves to earth, falling from bank to rills below, gurgling under gate-paths, lapping against the tree-trunks and little ridge piles in the brooks, and at last sweeping with a hushed content into the bosom of Thames. And the river himself was good for something more than a "stree-um." He was bank-full and sweeping on, taking to himself on this side and on that the tributes of his children, from which the waters poured so fast that they came in almost clear, and the mingled waters in the river were scarcely clouded in their flow. The lock-men rose by night and looked at the climbing flood, and wakened their wives and children, and raised in haste hatch after hatch of the weirs, and threw open locks and gates. Windsor Weir broke, but the wires flashed the news on, and the river's course was open, and after the greatest rain-storm and the lowest barometer known for thirty years, the Thames was not in flood, but only brimful; and once more a "river of waters."

THE SHELLS OF THE THAMES

Of the thousands who boat on the Thames during the summer few know or notice the beauty of the river shells. They are among the most delicate objects of natural ornament and design in this country. Exquisite pattern, graceful shapes, and in some cases lovely tints of colour adorn them. Nature has for once relaxed in their favour her rigid rules, by which she turns out things of this kind not only alike in shape, but with identical colour and ornament. Among humming-birds, for instance, each bird is like the other, literally to a feather. The lustre on each ruby throat or amethyst wing shines in the same light with the same prismatic divisions. But even in the London river, if you go and seek among the pebbles above Hammersmith Bridge when the river is low, you may find a score of *neretina* shells not one of which is coloured like the rest or ornamented with exactly the same pattern, yet each is fit to bejewel the coronet of some Titania of the waters. A number of these tiny shells, gathered from below the bridge, lie before the writer, set on black satin to display the hues. They look at a little distance like a series of mixed Venetian beads, but of more elegant form. From whichever side they are seen, the curves are the perfection of flowing line. The colouring and ornament of each is a marvel and delight. Some are black, with white spots arranged in lines following the curves, and with the top of the blunt spiral white. These "black-and-white marble" patterns are followed by a whole series in which purple takes the place of black, and the spots are modified into scales. Then comes a row of rose-coloured shells, some with white lance-heads, or scales, others with alternate bands of white scales and white dots. Some are polished, others dull, some rosy pink, others almost crimson. Some are marked with cream and purple like the juice of black currants with cream in it. In some the scale pattern changes to a chequer, some are white with purple zig-zags. And lastly come a whole series in pale olive, and olive and cream, in which the general colour is that of a blackcap's egg, and the pattern made by alternate spots of olive and bands of cream. If these little gems of beauty come out of the London river, what may we not expect in the upper waters of the silver Thames?[1] A search in the right places in its course will show. But these *neretinae* are everywhere up to the source of the river, for they feed on all kinds of decaying substances. If the pearl is the result of a disease or injury, the beauty of the *neretina* is a product or transformation from foul things to fair ones.

As the Thames is itself the product and union of all its vassal streams, an "incarnation" of all the rest, so in its bed it holds all the shells collected from all its tributaries. Different tribes of shells live in different waters. Some love the "full-fed river winding slow," some the swift and crystal chalk-stream. Some only flourish just over the spots where the springs come bubbling up from the inner cisterns of earth, and breathe, as it were, the freshness of these untainted waters; others love the rich, fat mud, others the sides of wearings and piles, others the river-jungles where the course is choked with weeds. But come what may, or flourish where they please, the empty shells are in time rolled down from trout-stream and chalk-stream, fountain and rill, mill-pool and ditch, cress-bed and water-cut, from the springs of the Cotswolds, the Chilterns, the downs, from the valleys of Berkshire, Buckinghamshire, Surrey, Gloucester, Oxford, and Essex, into the Thames. Once there the river makes shell collections on its own account, sorting them out from everything else except a bed of fine sand and gravel, in which they lie like birds' eggs in bran in a boy's cabinet, ready for who will to pick them up or sift them out of it. These shell collections are made in the time of winter floods, though how they are made or why the shells should all remain together, while sticks, stones, and other rubbish are carried away, it is impossible to say. They are laid on smooth points of land round which the waters flow in shallow ripples. Across the river it is always deep, swift, and dark, though the sandbanks come in places near the surface, and in the shallows grow water-crowfoot, with waving green hair under water, and white stems above it. The clean and shining sand shelves down to the water's edge, and continues below the surface. Here are living shells, or shells with living fish in them. In the bright water lie hundreds of the shells of the fresh-water mussels, the bearers of pearls sometimes, and always lined with that of which pearls are made, the lustrous nacre. The mealy masses of dry sand beyond the river's lip are stuffed with these mussel shells. They lie all ways up, endways, sideways, on their faces, on their backs. The pearl lining shines through the sand, and the mussels gleam like silver spoons under the water. They crack and crunch beneath your feet as you step across to search the mass for the smaller and rarer shells. Many of those in the water contain living mussels, yellow-looking fat molluscs, greatly beloved of otters, who eat them as sauce with the chub or bream they catch, and leave the broken shells of the one by the half-picked bones of the other. There was a popular song which had for chorus the question, "Did you ever see an oyster walk upstairs?" These mussels *walk*, and are said to be "tolerably active" by a great authority on their habits. They have one foot, on which they travel in search of feeding ground, and leave a visible track across the mud. There are three or four kinds, two of which sometimes hold small pearls, while a third is the pearl-bearer proper. *Unio pictorum* is the scientific name of one, because the shells were once the cups in which the old Dutch painters kept their colours, and are still used to hold ground gold and silver for illuminating. The pearl-bearing mussel is longer than the other kinds, flatter and darker, and the lining of mother-of-pearl is equal to half the total thickness of the shell.[2]

[Illustration: SHELLS OF THE THAMES. *From a photograph by E. Seeley*]

Though not so striking from their size and pearly lustre, there are many shells on the Thames sandbanks not less interesting and in large numbers. Among these are multitudes of tiny fresh-water cockle shells of all sizes, from that of a grain of mustard seed to the size of a walnut, flat, curled shells like small ammonites, fresh-water snail shells of all sizes, river limpets, *neretinae*, and other and rounder bivalve shells allied to the cockles. The so-called "snails" are really quite different from each other, some, the *paludinas*, being large, thick-striped shells, while the *limnaeas* are thin, more delicately made, some with fine, pointed spiral tops, and others in which the top seems to have been absorbed in the lower stories. There are eight varieties of these *limnaeas* alone, and six more elegant shells of much the same appearance, but of a different race.

The minute elegance of many of these shells is very striking. Tiny *physas* and *succineas*, no larger than shot, live among big *paludinas* as large as a garden snail, while all sizes of the larger varieties are found, from microscopic atoms to the perfect adult. Being water shells, and not such common objects as land shells, these have no popular names. The river limpets are called *ancylus fluviatilis*. Some are no larger than a yew berry, and are shaped like a Phrygian cap; but they "stick" with proper limpet-like tenacity. On the stems of water-lilies, on piles, on weeds and roots in any shallow streams, but always on the under side of the leaves, are the limpets of the Thames. The small ammonite-like shells are called *planorbis*, and like most of the others, belong also to the upper tertiary fossils. They feed on the decaying leaves of the iris and other water plants, and from the number of divisions on the shell are believed to live for sometimes twenty years. Of the many varieties, one, the largest, the horn-coloured *planorbis*, emits a purple dye. Two centuries ago Lister made several experiments in the hope that he might succeed in fixing this dye, as the Tyrians did that of the murex, but in vain. There are eleven varieties of this creature alone. It is easier to find the shells than to discover the living creature in the river. For many the deep, full river is not a suitable home; they only come there as the water does, from the tributary streams. Far up in some rill in the chalk, from the bed of which the water bubbles up and keeps the stones and gravel bright, whole beds of little pea-cockles may be found, lying in masses side by side, like seeds sown in the water-garden of a nymph.

[1] I have a series of *neretina* shells from the Philippines, much larger in size and brown in colour, in which many of the same kinds of ornament occur.

[2] A fresh-water mussel shell from North America in my possession is coloured green, and so marked and crimped as to resemble exactly a patch of water-weed, such as grows on stones and piles.

THE ANTIQUITY OF RIVER PLANTS

In the still gossamer weather of late October, when the webs lie sheeted on the flat green meadows and spools of the air-spiders' silk float over the waters, the birds and fish and insects and flowers of the best of England's rivers show themselves for the last time in that golden autumn sun, and make their bow to the audience before retiring for the year. All the living things become for a few brief hours happy and careless, drinking to the full the last drops of the mere joy of life before the advent of winter and rough weather. The bank flowers still show blossom among the seed-heads, and though the thick round rushes have turned to russet, the forget-me-not is still in flower; and though the water-lilies have all gone to the bottom again, and the swallows no longer skim over the surface, the river seems as rich in life as ever; and the birds and fish, unfrightened by the boat traffic, are tamer and more visible.

[Illustration: A FLOWERY BANK NEAR COOKHAM. *From a photograph by E.
Seeley.*]

The things in the waters and growing out of the waters are very, very old. The mountains have been burnt with fire; lava grown solid has turned to earth again and grows vines; chalk was once seashells; but the clouds and the rivers have altered not their substance. Also, so far as this planet goes, many of the water plants are world-encircling, growing just as they do here in the rivers of Siberia, in China, in Canada, and almost up to the Arctic Circle. The creatures which lived on these prehistoric plants live on them now, and in exactly the same parts of the stream. The same shells lie next the banks in the shallows as lie next the bank of the prehistoric river of two million years ago whose bed is cut through at Hordwell Cliffs on the Solent. The same shells lie next them in the deeper water, and the sedges and rushes are as "prehistoric" as any plant can well be. In the clay at Hordwell, which was once the mud of the river, lie sedges, pressed and dried as if in the leaves of a book, almost exactly similar in colour, which is kept, and in shape, which is uninjured, to those which fringe the banks of the Thames to-day. These freshwater plants show their hoary antiquity by the fashion of their generation. Most of them are mono-cotyledonous — with a single seed-lobe, like those of the early world. There is nothing quite as old among the Thames fishes as the mud fishes, the lineal descendants of the earliest of their race. But the same water creatures were feeding on the same plants perhaps when the Thames first flowed as a river.

[Illustration: BURR REED AND FLOWERING RUSH. *From photographs by E. Seeley.*]

The sedge fringe in the shallows, the "haunt of coot and tern" elsewhere, and of hosts of moorhens and dabchicks on the now protected river, is mainly composed of the giant rush, smooth and round, which the water-rats cut down and peel to eat the pith. These great rushes, sometimes ten feet high, *die* every year like the sickliest flowers, and break and are washed away. Few people have ever tried to reckon the number of kinds of sedges and reeds by the river, and it would be difficult to do so. There are forty-six kinds of sedge (*carex*), or if the *Scirpus* tribe be added, sixty-one, found in our islands. They are not all water plants, for the sand-sedge with its creeping roots grows on the sandhills, and some of the rarest are found on mountain-tops. But the river sedges and grasses, with long creeping roots of the same kind, have played a great part in the making of flat meadows and in the reclamation of marshes, stopping the water-borne mud as the sand-sedge stops the blowing sand. They have done much in this way on the Upper Thames, though not on the lower reaches of the river. The "sweet sedge," so called—the smell is rather sickly to most tastes—is now found on the Thames near Dorchester, and between Kingston and Teddington among other places, though it was once thought only to flourish on the Norfolk and Fen rivers. It is not a sedge at all, but related to the common arum, and its flower, like the top joints of the little finger, represents the "lords and ladies" of the hedges. So the burr reed, among the prettiest of all the upright plants growing out of the water, is not a reed, but a reed mace. Its bright green stems and leaves, and spiky balls, are found in every suitable river from Berkshire to the Amur, and in North America almost to the Arctic Circle. In the same way the yellow water villarsia, which though formerly only common near Oxford, has greatly increased on the Thames until its yellow stars are found as low as the Cardinal's Well at Hampton Court, extends across the rivers of Europe and Asia as far as China. The cosmopolitan ways of these water plants are easily explained. They live almost outside competition. They have not to take their chance with every new comer, for ninety-nine out of a hundred stranger seeds are quietly drowned in the embosoming stream. The water itself keeps its temperature steadily, and only changes slowly and in no great degree, and then, when the plants are in their winter sleep the stream may well say that "men may come and men may go, but I go on for ever." The same is very largely true of the things which live in the brook.

Many of the flowers are not quite what their names imply. The true lilies are among the oldest of plants. But "water-lilies" are not lilies. They have been placed in order between the barberry and the poppy, because the seed-head of a water-lily is like the poppy fruit. The villarsia, which looks like a water-lily, is not related at all, while the buck-bean is not a bean, but akin to the gentians. Water-violet might be more properly called water-primrose, for it is closely related to the primrose, though its colour is certainly violet, and not pale yellow. By this time all the bladderworts have disappeared under water. In June in a pool near the inflow of the Thames at Day's Lock, opposite Dorchester, the fine leafless yellow spikes of flower were standing out of the water like orchids, while the bladders with their trapdoors were employed in catching and devouring small tadpoles. There is something quietly horrible about these carnivorous plants. Their bladders are far too small to take one in whole, but catch the unhappy infant tadpoles by their tails and hold them till they die from exhaustion.

The bank flora of the Thames is nearly all the same from Oxford to Hampton Court, made up of some score of very fine and striking flowers that grow from foot to crest on the wall of light marl that forms the bank. Constantly refreshed by the adjacent water, they flower and seed, seed and flower, and are haunted by bees and butterflies till the November frosts. The most decorative of all are the spikes of purple loose-strife. In autumn when most of the flowers are dead the tip of the leaf at the heads of the spikes turns as crimson as a flower. The other red flowers are the valerian, in masses of squashed strawberry, and the fig-wort, tall, square-stemmed, and set with small carmine knots of flower. In autumn these become brown seed crockets, and are most decorative. The fourth tall flower is the flea-bane, and the fifth the great willow-herb. The lesser plants are the small willow-herbs, whose late blossoms are almost carmine, the water-mints, with mauve-grey flowers, and the comfrey, both purple and white. The dewberry, a blue-coloured more luscious bramble fruit, and tiny wild roses, grow on the marl-face also. At its foot are the two most beautiful flowers, though not the most effective, the small yellow snapdragon, or toad-flax, and the forget-me-not. This blue of the forget-me-nots is as peculiar as it is beautiful. It is not a common blue by any means, any more than the azure of the chalk-blue butterflies is common among other insects. Colour is a very constant feature in certain groups of flowers. One of these includes the forget-me-nots, the borage, the alkanet, and the viper's bugloss, which keep up this blue as a family heirloom. Others of the tribe, like the comfrey, have it not, but those which possess it keep it pure.

The willows at this time are ready to shed their leaves at the slightest touch of frost. Yet these leaves are covered with the warts made by the saw-flies to deposit their eggs in. The male saw-fly of this species and some others is scarcely ever seen, though the female is so common. The creature *stings* the leaf, dropping into the wound a portion of formic acid, and then lays its egg. The stung leaf swells, and makes the protecting gall. It is difficult to say when "fly," in the fisherman's use of the term as the adult insect food of fish, may not appear on the water. Moths are out on snowy nights, as every collector knows, and on any mild winter day flies and gnats are seen by streams. In the warm, sunny days of late September, numbers of some species of ephemerae were seen on the sedges and willows, with black bodies and gauzy wings, which the dace and bleak were swallowing eagerly, in quite summer fashion. The water is now unusually clear, and as the fish come to sun themselves in the shallows every shoal can be seen.

Among the typical Thames-valley flowers, all of which would be the better for protection, are the very rare soldier orchis (*Orchis Militaris*) and the monkey orchis (*Orchis Simia*), the water-snowflake, the *hottonia*, or water-violet, the water-villarsia, more elegant even than the water-lilies, the flowering rush, with a crown of bright rose-pink flowers. The two orchids named are very interesting plants. Of the monkey orchis Mr. Claridge Druce says in his "Flora of Oxfordshire" that it has become exceedingly scarce, not so much from the depredations of collectors, but from the fondness of rabbits for it and the changes brought about by agriculture. The soldier orchis is very rare indeed; both are only found in a few woods in the Thames valley, and possibly in Kent. The bladderworts fade instantly, and are not much interfered with, and though the fritillaries are picked for market, the roots are not dug up because that would injure the meadow turf in which they grow, and business objections would be raised.

INSECTS OF THE THAMES

Except among the select few, generally either enthusiastic boys or London mechanics of an inquiring mind, who keep fresh-water aquariums and replenish them from ponds and brooks at "weekends," few persons outside the fancy either see or know much of the water insects,[1] or are aware, when floating on a summer day under the willows in a Thames backwater, of the near presence of thousands of aquatic creatures, swift, carnivorous, and pursuing, or feeding greedily on the plants in the water garden that floats below the boat, or weaving nests, tending eggs, or undergoing the most astonishing transitions of form and activity, on or below the surface. Many of them are perhaps better equipped for encountering all the chances of existence than any other creatures. They can swim, dive, and run below water, live on dry land, or fly in the air, and many are so hardy as to be almost proof against any degree of cold. The great carnivorous water-beetle, the dytiscus, after catching and eating other creatures all day, with two-minute intervals to come up, poke the tips of its wings out of the water and jam some air against its spiracles, before descending once more to its subaqueous hunting-grounds, will rise by night from the surface of the Thames, lift again those horny wing-cases, unfold a broad and beautiful pair of gauzy wings, and whirl off on a visit of love and adventure to some distant pond, on to which it descends like a bullet from the air above. When people are sitting in a greenhouse at night with no lamp lighted, talking or smoking, they sometimes hear a smash as if a pebble had been dropped on the glass from above. It is a dytiscus beetle, whose compound eyes have mistaken the shine of the glass in the moonlight for the gleam of a pond. At night some of the whirligig beetles, the shiny, bean-like creatures seen whirling in incessant circles in corners by the bank, make a quite audible and almost musical sound upon the water. The activity of many of the water insects is astonishing. Besides keeping in almost incessant motion, those which spend most of their time below water have generally to come up constantly to breathe. Such are the water-bugs, water-scorpions and stick insects, which, though slender as rushes, and with limbs like hairs, can catch and kill the fry of the smaller fishes. Most of these are like divers, who have to provide themselves with air to breathe, and work at double speed in addition.

If a group of whirligig beetles is disturbed, the whole party will dive like dabchicks, rising to the surface again when they feel the need for breathing-air again. The diving-bell spiders, which do not often frequent the main Thames stream, though they are commonly found in the ditches near it, gather air to use just as a soldier might draw water and dispose it about his person in water-bottles. They do this in two ways, one of which is characteristic of many of the creatures which live both in and out of the water as the spider does. The tail of the spider is covered with black, velvety hair. Putting its tail out of the water, it collects much air in the interstices of the velvet. It then descends, when all this air, drawn down beneath the surface, collects into a single bubble, covering its tail and breathing holes like a coat of quicksilver. This supply the spider uses up when at work below, until it dwindles to a single speck, when it once more ascends and collects a fresh store. The writer has seen one of these spiders spin so many webs across the stems of water plants in a limited space that not only the small water-shrimps and larvae, but even a young fish were entangled. The other and more artistic means of gathering air employed by the spider is to catch a bubble on the surface and swim down below with it. The bubble is then let go into a bell woven under some plant, into which many other bubbles have been drawn. In this diving-bell the eggs are laid and the young hatched, under the constant watch of the old spider. Few people care to take the trouble to gaze for any time into a shallow, still piece of water, in which the bottom is plainly discernible, and a crop of water-weeds makes a wall on either side of some central "well." If they do find some such pond near the Thames banks or a shallow backwater, they may see after a few minutes much that is new and suggestive of strange activities. Everything will be quiet and motionless at first, for water beasts are very suspicious of movement above them, and all sham dead, or lie quite still, and are strangely invisible. On the other hand, they have none of the power of remaining motionless for half-an-hour like land animals. Soon what look like sticks, but are caddis larva, begin to creep on the bottom. Then more brown objects, larvae of dragon-flies and water-beetles, detach themselves from the stems of the plants and cruise up and down seeking what they may devour. Other creatures feeding and swimming among or beneath the plants crawl out on to the upper surface, and the water-beetles come up to breathe, or to play upon the surface. One of the largest of these is a very fine *black* beetle, a vegetable-feeding creature. It is most interesting to see two of them—they generally live in pairs—browsing on one of the fern-like plants of the Thames. This plant has leaves like fern blades, each having in turn its own small spikelets. The big beetles work along the leaf like a cow in a cabbage yard, biting off, chewing, and swallowing each in succession, and leaving the stem perfectly bare. Sometimes it looks as if the two beetles were eating for a match, like the beef-eating contests held in country public-houses, in which the winner once boasted that he won easily "afore he came to vinegar."

The number of carnivorous creatures found in the water seems out of all proportion to the usual order of Nature. But this is perhaps because the minute, almost invisible creatures, or entomostraca, of which the rivers and ponds are full, and which are the main food of the smaller water carnivora, live mainly on decaying vegetable substance, which is practically converted and condensed into microscopical animals before these become in turn the food of others. It is as if all trees and grass on land were first eaten by locusts or white ants, and the locusts and white ants were then eaten by semi-carnivorous cows and sheep, which were in turn eaten by true carnivora. The water-weeds, both when living and decaying, are eaten by the entomostraca, the entomostraca are eaten by the larvae of insects, the perfect insects are eaten by the fish, and the fish are eaten by men, otters, and birds. Thus we eat the products of the water plants at four removes in a fish; while we eat that of the grass or turnips only in a secondary form in beef or mutton.

The water-shrimp is a very common crustacean in the small Thames tributaries, and valuable as fish food. It has a very rare subterranean cousin known as the *well shrimp*. A lady in the Isle of Wight, who in a moment of energy went to the pump to get some water to put flowers in, actually pumped up one of these subterranean shrimps into a glass bowl. The well was eighty feet deep. The shrimp was absolutely white, and probably blind.

Flesh-eating insects are fairly common on land; wasps will actually raid a butcher's shop, and carry off little red bits of meat, besides killing and eating flies, spiders, and larvae. Dragon-flies are the hawks of the insect world, and slay and devour wholesale, when in the air as well as when they are larvae on the water, though few persons actually witness their attacks on other creatures, owing to the swiftness of their flight. Some centipedes will attack other creatures with the ferocity of a bulldog. An encounter between one of the smaller centipedes and a worm is like a fight between a ferret and a snake, so frantic is the writhing of the worm, so determined the hold which the hard and shiny centipede maintains with its hooked jaws. But the ferocity and destroying appetite of some of the water creatures would be appalling were it not for their small size. The desire of killing and devouring appears in the most unexpected quarters, among creatures which no one would suspect of such intentions. Of two kinds of water snail found in the Thames, and among the commonest molluscs, one is a vegetable feeder. It is found living on water plants, the snails being of all sizes, from that of a mustard seed to a walnut. The other will feed not only on dead animal substances, but on living creatures, and is equipped with sharp teeth, which work like a saw. One of these kept in an aquarium fastened on to and slowly devoured a small frog confined in the same vessel. The large dytiscus beetle is the great enemy of small fish. If the salmon is ever restored to the Thames these creatures will be among the worst enemies of the fry, though in swift rivers they are not plentiful. Frank Buckland states that in Hollymount Pond they killed two thousand young salmon. One of these was put into a bowl with a dytiscus beetle, which, "pouncing upon him like a hawk upon an unsuspecting lark, drove its scythe-like horny jaws right into the back of the poor little fish. The little salmon, a plucky fellow, fought hard for his life, and swam round and round, up and down, hither and thither, trying to escape from this terrible murderer; but it was no use, he could not free himself from his grip; and while the poor little wretch was giving the last few flutterings of his tail, the water-beetle proceeded coolly to peck out his left eye, and to devour it at once." The larva not only of the carnivorous dytiscus but also of the vegetable-feeding water-beetle are ferocious and carnivorous, and deadly enemies of young fish and ova.

[1] In mentioning some of the Thames *insecta* I have also noticed some of the *mollusca* and *crustacea*. It is a pity these have not some common names. One cannot write easily of "pulmonate gasteropods."

"THE CHAVENDER OR CHUB"

"Now when you've caught your chavender,
  (Your chavender or chub)
You hie you to your pavender,
  (Your pavender or pub),
And there you lie in lavender,
  (Sweet lavender or lub)."

Mr. Punch.

I went into the Plough Inn at Long Wittenham in mid-November to arrange about sending some game to London. The landlord, after inquiring about our shooting luck, went out and came back into the parlour, saying, "Now, sir, will you look at my sport?" He carried on a tray two large chub weighing about 2-1/2 lbs. each, which he had caught in the river just behind the house. Their colour, olive and silver, scarlet, and grey, was simply splendid. Laid on the table with one or two hares and cock pheasants and a few brace of partridges they made a fine sporting group in still life—a regular Thames Valley yield of fish and fowl. The landlord is a quiet enthusiast in this Thames fishing. It is a pleasure to watch him at work, whether being rowed down on a hot summer day by one of his men, and casting a long line under the willows for chub, or hauling out big perch or barbel. All his tackle is exquisitely kept, as well kept as the yeoman's arrows and bow in the Canterbury Tales. His baits are arranged on the hook as neatly as a good cook sends up a boned quail. He gets all his worms from Nottingham. I notice that among anglers the man who gets his worms from Nottingham is as much a connoisseur as the man who imported his own wine used to be among dinner-givers.

Drifting against a willow bush one day, the branches of which came right down over the water like a crinoline, I saw inside, and under the branches, a number of fair-sized chub of about 1 lb. or 1-1/2 lbs. It struck me that they felt themselves absolutely safe there, and that if in any way I could get a bait over them they might take it. The entry under which I find this chronicled is August 24th. Next morning when the sun was hot I got a stiff rod and caught a few grasshoppers. Overnight I had cut out a bough or two at the back of the willow bush, and there was just a chance that I might be able to poke my rod in and drop the grasshopper on the water. After that I must trust to the strength of the gut, for the fish would be unplayable. It was almost like fishing in a faggot-stack. Peering through the willow leaves I could just see down into the water where a patch of sunlight about a yard square struck the surface. Under this skylight I saw the backs of several chub pass as they cruised slowly up and down. I twisted the last two feet of my line round the rod-top, poked this into the bush with infinite bother and pluckings at my line between the rings, and managed to drop the hopper on to the little bit of sunny water. What a commotion there was. The chub thought they were all in a sanctuary and that no one was looking. I could see six or seven of them, evidently all cronies and old acquaintances, the sort of fish that have known one another for years and would call each other by their Christian names. They were as cocky and consequential as possible, cruising up and down with an air, and staring at each other and out through the screen of leaves between them and the river, and every now and then taking something off a leaf and spitting it out again in a very independent connoisseur-like way. The moment the grasshopper fell there was a regular rush to the place, very different from what their behaviour would have been outside the bush. There was a hustle and jostle to look at it, and then to get it. They almost fought one another to get a place. Flop! Splash! Wallop! "My grasshopper, I think." "I saw it first." "Where are you shoving to?" "O—oh—what is the matter with William?" I called him William because he had a mark like a W on his back. But he was hooked fast and flopping, and held quite tight by a very strong hook and gut, like a bull with a ring and a pole fastened to his nose. I got him out too—not a big fish, but about 1-1/2 lbs.

This showed pretty clearly that where chub can be fished for "silently, invisibly," they can still be caught, even though steam launches or row-boats are passing every ten minutes. This was mid-August; my next venture nearly realised the highest ambitions of a chub-fisher. It also showed the sad limitations of mere instinctive fishing aptitudes in the human being as contrasted with the mental and bodily resources of a fish with a deplorably low facial angle and a very poor *morale*. There was just one place on the river where it seemed possible to remain unseen yet to be able to drop a bait over a chub. A willow tree had fallen, and smashed through a willow *bush*. Its head stuck out like a feather brush in front and made a good screen. On either side were the boughs of the bush, high, but not too high to get a rod over them, if I walked along the horizontal stem of the tree. It was only a small tree, and a most unpleasant platform. But I had caught a most appetising young frog, rather larger than a domino, which I fastened to the hook, and after much manoeuvring I dropped this where I knew some large chub lay. As the tree had only been blown down a day before, I was certain that they had never been fished for at that spot.

[Illustration: A MONSTER CHUB. *From a drawing by Lancelot Speed.*]

I was right; hardly had the frog touched the water when I saw a monster chub rise like a dark salamander out of the depths. Slowly he rose and eyed the frog, moving his white lips as if the very sight imparted a gusto to the natural excellence of young frogs. I nearly dropped from the tree stem from sheer suspense, when he made up his mind, put on steam, and took it! He was fast in a minute, and kindly rushed out into the river, where I played him. Then I wound in my line and hauled him up till his head and mouth were out of the water. As there was an impenetrable screen of bushes between him and me I laid the rod down, trusting to the tackle, and ran round to where close by was a farm punt, made fast. It had been used during harvest time, and was full of what in the classics they call the "implements of Ceres." All of these that do not seem made to cut your leg off are designed to run into and spike you. Besides scythes and reap hooks, there were iron rakes (sharp end upwards), wooden rakes, pitchforks, and garden forks, and the difficulty was to move in the punt without getting cut or spiked. The last users of the punt had also taken peculiar care to fasten it up. It was anchored by a grapnel, and by an iron pin on a chain, the pin eighteen inches long and driven hard into the bank. In a desperate hurry I hauled up the grapnel, did a regular Sandow feat in pulling up the iron peg, seized a punt pole apparently weighted with lead, but made out of an ash sapling, and started the punt. It would not move. I found there was another mooring, so picking my way among the scythes, spikes, rakes, &c., I hauled this in. It was most infernally heavy, and turned out to be a cast-iron wheel of a steam plough or other farming implement. Then I was under weigh, and got round to the fish. It was still there. I could see its expressionless eye (about as big as a sixpence) out of the water and its mouth wide open, when I remembered I had forgotten the landing-net in my hurry. Then came the period of mental aberration common to the amateur. The fish was certainly 4 lbs. in weight, yet I tried to get him in with my hands. Of course he gave one big flop, slipped out, and disappeared — the biggest chub I ever shall not catch.

THE WORLD'S FIRST BUTTERFLIES

Thames plants must strike every one as belonging to an ancient order of life. But the vast clouds of winged *ephemeridae* that dance over its waters when there is a rise of "May-fly" in early summer look to be not only the creatures of a day, but of our day. In the astonishing wave and rush of life seen at such times, when from every plant and pool winged creatures are ascending to float in air, it is difficult to picture the silence and stillness of a world where there were no birds, or hum of bees, and no signs of the other insects which exceed the other population of the earth by unnumbered myriads of millions; yet the insects, even the same identical species which dance over the Thames to-day, are among the very oldest of living things, just as its plants and its shells are. Rocks and slate are not ideal butterfly cases; and if the fragile limbs of the beetle and grasshopper of the successive prehistoric worlds had perished beyond the power of identification, no one could have felt surprise. But such has been the industry of modern naturalists—to give the widest name to those who have devoted their time to the search for, and description of, fossil insects—that the remains of thousands of species have been identified, and the time of their appearance upon the earth approximately fixed. The latest contributor to this elegant branch of the study of fossils is Mr. Herbert Goss.[1] Perhaps the most interesting of his conclusions is the antiquity, not only of the existing orders of insects, but even of their particular families and genera, as compared with vertebrate animals. It is astonishing to find not only crickets and beetles existing at periods enormously earlier than the appearance of birds or fish, but that they conformed in type to the families in which they are classed to-day. Though they become fewer and fewer as they are tracked back up the river of time, there are not found in the earliest fossil-bearing rocks any connecting links or earlier and simpler forms of insect life, or a clue to the common ancestor of insects, spiders, and shrimps, which naturalists would dearly like to discover. There is a baffling completeness about these creatures. When in the lias period, for instance, the vertebrates were huge saurian reptiles and flying lizards, and scarcely any of our existing classes of fish had come into existence, the beetles, cockroaches, crickets, and white ants were there, with all the distinguishing characteristics of the existing families as they were settled by Linnaeus.

The first insect known to have existed, a creature of such vast antiquity that it deserves all the respect which the parvenu man can summon and offer to it, was—a cockroach. This, the father of all black-beetles, probably walked the earth in solitary magnificence when not only kitchens, but even kitchen-middens were undreamt of, possibly millions of years before Neolithic man had even a back cave to offer with the remains of last night's supper for the cockroach of the period to enjoy. His discovery established the fact that in the Silurian period there were insects, though, as the only piece of his remains found was a wing, there has been room for dispute as to the exact species. Mr. Goss in his preface to the second edition of his book notes that what is probably a still older insect has been found in the lower Silurian in Sweden. This was not a cockroach, but apparently something worse. If the Latin name, *Protocimex Silurius*, be literally translated, it means the original Silurian bug. It was a fair conjecture that insects appeared about the same time as land plants first grew on the earth. As almost all the species either feed on some vegetable substances in growth or decay, or else live upon other insects, some such provision of food was necessary for them. Remains of such plants were discovered in the Silurian rocks. In the Devonian formations, which contain the next oldest set of fossil insects, numbers of conifers and ferns are found. Yet even then the only vertebrate animals seem to have been fish. The insects still had the land all to themselves. Of one of these Devonian insects the base of a wing was the only part preserved in the rock. From this it was possible to tell the order to which the creature belonged. It was one of the *Neuroptera* —insects with wings in which the veins run straight down the wing, sometimes joined by cross branches at right angles. Some of the modern kinds are very beautiful four-winged flies, with bright colours on their wings like butterflies. Others are ant-lions or caddis-flies. The curve of the fragment of wing also suggested its probable size when unbroken. It was perhaps two inches long. As there are little horny rings round the wing base like those which crickets have, on which they rub their legs and so "chirp," it is also quite likely that this insect of hoary antiquity did the same, and enlivened the silence of Devonian fern groves with a prehistoric hum. It is quite in keeping with modern ideas that in that age of fishes one of the most remarkable insects should have been a kind of May-fly, "a large species of *Ephemerina*, which must have measured five inches in expanse of wings." Thus our Thames May-flies had gigantic prehistoric ancestors, which appeared on earth, possibly with their present associates the caddis flies, at an enormously remote age.

So far no butterfly had yet appeared on earth, though the *Ephemerinae* might dance over the still lagoons and swamps. In the coal-forest period, and the age of trees and rank vegetation, insects of many kinds seem to have multiplied, even though the most beautiful of all were not yet launched in air. In England the first beetle wandered on to the stage of life—the oldest British insect fossil known. It was discovered in the ironstone of Coalbrookdale, and was a kind of weevil. Another creature found in the same ironstone was a cricket. It is quite in keeping with the forest and tree surroundings of the time that white ants should have abounded to eat up the decayed and dead wood. Strictly speaking, black-beetles are not beetles at all. But they are a very good imitation. As some hundreds of families of *Paltaeoblattidae*, which may be translated as "old original cockroaches," and *Blattidae*, or cockroaches *pur sang*, pervaded these forests, and the doyen of all Swiss fossil animals is one of these, the "state of the streets" in a coal forest may be imagined when there were no bird police to keep the insects in order. Thus the end of the Palaeozoic world—a very poor world at best—was fairly well stocked with insects, though the moths, bees, and butterflies had yet to come. Then came the sunrise of a new time—mammals, any number of reptiles, possibly some birds, and an insect life more teeming than any we now know. The "insect limestone" attests these multitudes. Beetles, of which the scarabs were a numerous family, increased vastly, and the oldest known dragon-fly and supposed ancestor of those which hawk over the Oxford river, left his skeleton, or what represents a dragon-fly's skeleton, among some two thousand other specimens of fossil insects, in the Swiss Alps. It was then that the first bird and the first butterfly appeared. The bird was the famous Archaeopteryx, found in the Solenhofen slate, and the first butterfly, to use an Irishism, was a moth, a sphinx moth, apparently about the size of the Convolvulus sphinx moth. This stone-embedded relic of the moth that sucked the juices of the plants of the Mesozoic world, incalculable ages before the time even of the gigantic mammals, is preserved in the Teyler Museum at Haarlem. When the new era of the Eocene period developed modern forms of plants, their rapid growth was accompanied by a great increase in the number of insects. Those which, like the moths, had only made their first venture on earth, now appeared in greater numbers. Near Aix, in Provence, five butterflies and two moths were found in some beds of marl and gypsum long celebrated for their fossils, and with the fossil butterflies were, in every case but one, fossil remains of the plants which had served its larvae as food. Thus the May-flies and beetles are perhaps older than the Thames shells, and older than the prehistoric plants on which the river molluscs feed.

[1] Secretary of the Entomological Society, and an accomplished botanist. The work is entitled "The Geological Antiquity of Insects," and published by Gurney and Jackson, London.

BUTTERFLY SLEEP

Fond as the butterflies are of the light and sun, they dearly love their beds. Like most fashionable people who do nothing, they stay there very late. But their unwillingness to get up in the morning is equalled by their equal desire to leave the world and its pleasures early and be asleep in good time. They are the first of all our creatures to seek repose. An August day has about fifteen hours of light, and for that time the sun shines for twelve hours at least; but the butterflies weary of sun and flowers, colour and light, so early that by six o'clock, even on warm days, many of them have retired for the night. I climbed Sinodun Hill, on a cold, windy afternoon, and found that hundreds of butterflies were all falling asleep at five o'clock. Their dormitory was in the tall, colourless grass, with dead seed-heads, that fringes the tracks over the hills, or the lanes that cross the hollows. Common blues were there in numbers, and small heath butterflies almost as many. The former, each and every one of them, arrange themselves to look like part of the seed-spike that caps the grass-stem. Then the use and purpose of the parti-coloured grey and yellow under-colouring of their wings is seen. The butterfly invariably goes to sleep head downwards, its eyes looking straight down the stem of the grass. It folds and contracts its wings to the utmost, partly, perhaps, to wrap its body from the cold. But the effect is to reduce its size and shape to a narrow ridge, making an acute angle with the grass-stem, hardly distinguishable in shape and colour from the seed-heads on thousands of other stems around.[1] The butterfly also sleeps on the top of the stem, which increases its likeness to the natural finial of the grass. In the morning, when the sunbeams warm them, all these grey-pied sleepers on the grass-tops open their wings, and the colourless bennets are starred with a thousand living flowers of purest azure. Side by side with the "blues" sleep the common "small heaths." They use the grass-stems for beds, but less carefully, and with no such obvious solicitude to compose their limbs in harmony with the lines of the plant. They also sleep with their heads downwards, but the body is allowed to droop sideways from the stem like a leaf. This, with their light colouring, makes them far more conspicuous than the blues. Moreover, as grass has no leaves shaped in any way like the sleeping butterfly, the contrast of shape attracts notice. Can it be that the blues, whose brilliant colouring by day makes them conspicuous to every enemy, have learnt caution, while the brown heaths, less exposed to risk, are less careful of concealment? Be it noticed that moths and butterflies go to sleep in different attitudes. Moths fold their wings back upon their bodies, covering the lower wing, which is usually bright in colour, with the upper wing. They fold their antennas back on the line of their wings. Butterflies raise the wings above their bodies and lay them back to back, putting their antennae between them if they move them at all. On these same dry grasses of the hills, another of the most brilliant insects of this country may often be seen sleeping in swarms — the carmine and green burnet moth. But it is a sluggish creature, which often seems scarcely awake in the day, and its surrender to the dominion of sleep excites less surprise than the deep slumber of the active and vivacious butterflies. The "heaths" and "blues" should perhaps be regarded as the gipsies of the butterfly world, because they sleep in the open.

They are even worse off than the nomads, because, like that regiment sleeping in the open which the War Office lately refused to grant field allowance to on the ground that they were "not under canvas," they do not possess even a temporary roof. What we may call the "garden butterflies," especially the red admirals, often do seek a roof, going into barns, sheds, churches, verandahs, and even houses to sleep. There, too, they sometimes wake up in winter from their long hibernating sleep, and remind us of summer days gone by as they flicker on the sun-warmed panes. Mrs. Brightwen established the fact that they sometimes have fixed homes to which they return. Two butterflies, one a brimstone, the other, so far as the writer remembers, a red admiral, regularly came for admission to the house. One was killed by a rain-storm when the window was shut; the other hibernated in the house. Probably it was as a sleeping-place and bedroom that the butterflies made it their home. There is a parallel instance, mentioned by a Dutch naturalist quoted by Mr. Kirby, when a butterfly came night after night to sleep on a particular spot in the roof of a verandah in the Eastern Archipelago. In the East the sun itself is so regular and so rapid in rising and setting that the sleeping hours of insects and birds are far more regular than in temperate lands, with their shifting periods of light and darkness. Our twilight, that season that the tropics know not, has produced a curious race of moths, or rather, a curious habit confined to certain kinds. They are the creatures neither of day nor of night, but of twilight. They awake as twilight begins, go about their business and enjoy a brief and crepuscular activity, and go to sleep as soon as darkness settles on the world. At the first glimmer of the dawn they awaken again to fly till sunrise, when they hurry off like the fairies, and sleep till twilight falls again.

[Illustration: BUTTERFLIES AT REST. *From photographs by R.B. Lodge.*]

At the time of writing a border of bright flowers runs in straight perspective from the window opposite, with a rose arcade by the border, and a yew hedge behind that. The shafts of the morning sun fly straight down to the flowers, and every blossom of hollyhock, sunflower, campanula, and convolvulus, and the scarlet ranks of the geraniums, are standing at "attention" to welcome this morning inspection by the ruler and commander-in-chief of all the world of flowers. The inspecting officers, rather late as inspecting officers are wont to be, are overhauling and examining the flowers. These inspectors, also roused by the sun, are the butterflies and bees. Splendid red admirals are flying up, and alighting on the sunflowers, or hovering over the pink masses of valerian. Peacock butterflies, "eyed" like Emperors' robes, open and shut their wings upon the petals; large tortoiseshells are flitting from flower to flower; mouse-coloured humming-bird moths are poising before the red lips of the geraniums; and a stream of common white butterflies is crossing the lawn to the flowers at the rate of twenty a minute. They all come from the same direction, across a cornfield and meadow, behind which lies a wood. The bees came first, as they are fairly early risers; the butterflies later, some of them very late, and evidently not really ready for parade, for they are sitting on the flowers stretching, brushing themselves, and cleaning their boots—or feet. The fact is that the butterflies, late though it is, are only just out of bed. You might look all the evening to find the place where these particular butterflies sleep, and not discover it, unless some of them have taken a fancy to the verandah or the inside of a dwelling-room in the house. But each and every one of them has been asleep in a place it has chosen, and it is probable that some, the red admirals, for instance, will go back to that place to sleep at evening.

As there are hundreds of moths that fly by night and sleep by day at seasons when there are perhaps only twenty species of butterflies flying by day and sleeping by night, it is strange that the sleeping moths are not more often found. Some kinds are often disturbed, and are seen. But the great majority are sleeping on the bark of trees, in hedges, in the crevices of pines, oaks and elms, and other rough-skinned timber, and we see them not. Some prefer damp nights with a drizzle of rain to fly in, not the weather which we should choose as inviting us to leave repose. Few like moonlight nights; darkness is their idea of a "fine day" in which to get up and enjoy life, many, like the dreams in Virgil's Hades, being all day high among the leaves of lofty trees, whence they descend at the summons of night, the—

"Filmy shapes
That haunt the dusk, with ermine capes,

And woolly breasts, and beaded eyes,"

The connection between character and bedtime which grew up from association when human life was less complex than now has some counterpart in the world of butterflies and insects. The industrious bees go to bed much earlier than the roving wasps. The latter, which have been out stealing fruit and meat, and foraging on their own individual account, "knock in" at all hours till dark, and may sometimes be seen in a state of disgraceful intoxication, hardly able to find the way in at their own front door. The bees are all asleep by then in their communal dormitory.

It would not be human if some belief had not arisen that the insects that fly by night imitate human thieves and rob those which toil by day. There has always been a tradition that the death's-head moth, the largest of all our moths, does this, and that it creeps into the hives and robs the bees, which are said to be terrified by a squeaking noise made by the gigantic moth, which to a bee must appear as the roc did to its victims. It is said that the bees will close up the sides of the entrance to the hive with wax, so as to make it too small for the moth to creep in. Probably this is a fable, due to the pirate badge which the moth bears on its head. But it is certainly fond of sweet things, and as it is often caught in empty sugar-barrels, it is quite possible that it does come to the hive-door at night and alarm the inmates in its search for honey.

[1] In the illustration it was impossible to photograph butterflies actually sleeping. They show their attitude, but not the degree to which the wings are flattened into a very acute angle.

CRAYFISH AND TROUT

About the middle of August, when walking by one of the locks on a disused canal in the Ock Valley, I saw a man engaged in a very artistic mode of catching crayfish. The lock was very old, and the brickwork above water covered with pennywort and crane's-bill growing where the mortar had rotted at the joints. In these same joints below water the crayfish had made holes or homes of some sort, and were sitting at the doors with their claws and feelers just outside, waiting, like Mr. Micawber, for something to turn up. To meet their views the crayfish catcher had cut a long willow withe. From the tapering tip of this he had cut the wood, leaving the bark, which had been carefully slit and the woody tip extracted from it. This pendant of bark he had made into a running noose, and leaning over the bank he worked it over the crayfish's claws and then snared them. It was a neat adaptation of local means to an end; for if you think of it, string would not have answered, because it would not remain rigid, and wire would be too stiff for the job.

Crayfish catching, until lately one of the minor fisheries of the Thames, is now a vanished industry. Ten years ago the banks of the river from Staines to the upper waters at Cricklade were honeycombed with crayfish holes, like sandmartins' nests in a railway cutting. These holes were generally not more than eighteen inches below the normal water line of the river. In winter when the stream was full fresh holes were dug higher up the bank. In summer when the water fell these were deserted. The result was that there were many times more holes than crayfish, and that for hundreds of miles along the Thames and its tributaries these burrows made a perforated border of about three feet deep. The almost complete destruction of the crayfish was due to a disease, which first appeared near Staines, and worked its way up the Thames, with as much method as enteric fever worked its way down the Nile in the Egyptian Campaign after Omdurman. The epidemic is well known in France, where a larger kind of crayfish is reared artificially in ponds, and serves as the material for *bisque d'écrevisses*, and as the most elegant scarlet garnish for cold and hot dishes of fish in Paris restaurants; but it was new to recent experience of the Thames. Perhaps that is why its effects were so disastrous. The neat little freshwater lobsters turned almost as red as if they had been boiled, crawled out of their holes, and died. Under some of the most closely perforated banks they lay like a red fringe along the riverside under the water. Near Oxford, and up the Cherwell, Windrush, and other streams they were, before the pestilence, so numerous that making crayfish pots was as much a local industry as making eel-pots, the smaller withes, not much larger than a thick straw, being used for this purpose. Most cottages near the river had one or two of these pots, which were baited on summer nights and laid in the bottom of the stream near the crayfish holes. It must be supposed that they only use them by day, and come out by night, just as lobsters do, to roam about and seek food on a larger scale than that which they seize as it floats past their holes by day. That time of more or less enforced idleness the crayfish used to spend in looking out of their holes with their claws hanging just over the edge ready to seize and haul in anything nice that floated by. Their appetite by night was such that no form of animal food came amiss to them. The "pots" were baited with most unpleasant dainties, but nasty as these were they were not so unsavoury as the food which the crayfish found for themselves and thoroughly enjoyed, such as dead water-rats and dead fish, worms, snails, and larvae. They were always hungry, and one of the simplest ways of catching them was to push into their holes a gloved finger, which the creature always seized with its claw and tried to drag further in. The crayfish, who, like the lobster, looked on it as a point of honour never to let go, was then jerked out into a basket. They rather liked the neighbourhood of towns and villages because plenty of dirty refuse was thrown into the water. In the canalised stream which runs into Oxford city itself there were numbers, which not only burrowed in the bank, but made homes in all the chinks of stone and brick river walls, and sides of locks, and in the wood of the weiring, where they sat ensconced as snugly as crickets round a brick farmhouse kitchen fireplace. They were regularly caught by the families of the riverine population of boatmen, bargees, and waterside labourers, and sold in the Oxford market. A dish of crayfish, as scarlet as coral, was not unfrequently seen at a College luncheon. Possibly the recovery from the epidemic may be rapid, and the small boys of Medley and Mill Street may earn their sixpence a dozen as delightfully as they used to. Young crayfish, when hatched from the egg, are almost exactly like their parents. The female nurses and protects them, carrying them attached to its underside in clinging crowds. They grow very fast, and this makes it necessary for the youthful crayfish to "moult" or shed their shells eight times in their first twelvemonth of life, as the shell is rigid and does not grow with the body. The constant secretion of the lime necessary to make these shells is so exhausting to the youthful crayfish that only a small number ever grow up. In America, where a large freshwater crayfish nearly a foot long is found, its burrowing habits are a serious nuisance, especially in the dykes of the Mississippi. In those streams from which these interesting little creatures have entirely disappeared it might be worth while to introduce the large Continental crayfish. As it is bred artificially, there would be no difficulty in obtaining a supply, and it would be a useful substitute for the small native kind.

Sea crayfish, which grow to a very large size, are not much esteemed in this country. They are not so well flavoured as their cousin the lobster. But as river crayfish of a superior kind can be cultivated, and are reared for the table abroad, it might be worth while to pay some attention to what has been done in the United States to replenish by artificial breeding the stock of lobsters now somewhat depleted by the great "canning" industry. The method of obtaining the young lobsters is different from that employed to rear trout from ova. The female lobsters carry all their eggs fastened to hair-fringed fans or "swimmerets" under their tails, the eggs being glued to these hairs by a kind of gum which instantly hardens when it touches the water. For some ten months the female lobster carries the eggs in this way, aerating them all the time with the movement of the swimmerets. When they are caught in the lobster-pots in the months of June and July, the eggs are taken to the hatchery, and the ova are detached. As they are already fertilised, they are put into hatching jars, where in due course they become young lobsters, or rather lobster larvae, for the lobster does not start in life quite so much developed as does the infant crayfish. It is about one-third of an inch long, has no large claws, and swims naturally on the surface of the water, instead of lurking at the bottom as it does when it has come to lobster's estate. It seems to be compelled to rise to the surface, for sunlight, or any bright illumination, always brings swarms of lobsterlings to the top of the jars in which they are hatched. In the sea this impulse towards the light stands them in good stead, for in the surface-waters they find themselves surrounded by the countless atoms of animal life, or potential life, the eggs and young of smaller sea beasts. The young lobster is furiously hungry and voracious, because, like the young crayfish, it has to change not only its shell but the lining of its stomach five times in eighteen days. Unfortunately, in the hatching jars there is no such store of natural food as in the sea. The result is that the young lobsters have to eat each other, which they do with a cheerful mind, if they are not at once liberated. When they have reached their fifth month they go to the bottom and "settle down" in the literal sense to the serious life of lobsters.

[Illustration: A TROUT. *From a photograph by Charles Reid.*]

I believe no one ever saw trout spawning in the Thames, though there are plenty of shallows where they might do so. Consequently the Thames trout must be regarded as a fish which was born in the tributaries and descended into the big river, and as the mouths of these trout-holding tributaries, such as the Kennet at Reading, the Pang, the lower Colne, and others, become surrounded with houses and the trout no longer haunt the *embouchure*, so the tendency is for fewer trout to get into the Thames. Still, places like the Windrush, the Evenlode, and the other upper tributaries hold rather more trout than they did, as they are better looked after; and the Fairford Colne is still a beautiful trout stream. For some reason, however, the Thames trout do not seem fond of the upper waters, where if found they seem to keep entirely in the highly aerated parts by the weirs, but mainly haunt the lower ones from Windsor downwards, and one was recently caught in the tidal waters below the bridge. It is very difficult to see why there are so few above Oxford, or from Abingdon to Reading. It is not because they are caught, for very few are caught. A friend of mine who had lived on the river near Clifton Hampden for some eight years, could only remember eight trout being caught in that time. I thought I was going to have one once. I was fishing for chub with a bumble bee, and a great spotted trout rose to it in a way which made me hope I was going to have a trophy to boast of for life. But he "rose short," and I saw him no more. I believe *all* the brooks which rise in the chalk hills of the Thames Valley have trout in them. One runs under the railway line at Steventon. A resident there had quite a number of tamed trout in the conduit which took the stream under the line, and used to feed them with worms as a show. At the head waters of the Lockinge brook, close to the springs, I saw the trout spawning on New Year's Day. The big fish had wriggled up into the very shallowest water, and were lying with their back fins and tails out, I suppose from some instinct either that this water is the most highly aerated, or because floods do less harm on a shallow, or for both reasons combined. At Long Wittenham, though I never saw a trout in the river (they are, however, taken there), Admiral Clutterbuck recently had a fine old stew pond in the picturesque old grounds of the Manor House cleaned out, and stocked it with rainbow trout. They did well and grew fast, and so far as I know, none died. The water was not suited for their breeding, but the fish were very ornamental, and rose freely to the fly.

FOUNTAINS AND SPRINGS

Is it true that our fountains and springs of sweet water are about to perish? A writer in *Country Life* says "Yes," that in parts of the Southern counties the hidden cisterns of the springs are now sucked dry, and that the engineers employed to bring the waters from these natural sources to the village or the farm lament that where formerly streams gushed out unbidden, they are now at pains to raise the needed water by all the resources of modern machinery. When the old fountains fail new sources are eagerly sought, and where science fails the diviner's art is called in to aid. At the Agricultural Show the water-diviner sits installed, surrounded by votive tablets picturing the springs discovered by his magic art; and County Councils quarrel with the auditors of local expenditure over sums paid for the successful employment of his mysterious gift.

It is not strange that the springs of England should still suggest a faint echo of Nature-worship. If rivers have their gods, fountains and springs have ever been held to be the home of divinities, beings who were by right of birth gods, even though, owing to circumstances, they did not move exactly in their circle. *Procul a Jove, procul a fulgure* may have been the thought ascribed by Greek fancy to the gracious beings who made their home by the springs, for whether in ancient Greece or in our Western island, they breathe the sense of peace, security, and quiet, and to them all living things, animal and human, come by instinct to enjoy the sense of refreshment and repose. A spring is always old and always new. It is ever in movement, yet constant, seldom greater and seldom less, in the case of most natural upspringing waters, syphoned from the deep cisterns of earth. Absolutely material, with no mystery in its origin, it impresses the fancy as a thing unaccountable, like the source of life embodied, something self-engendered. It has pulses, throbbing like the ebb and flow of blood. Its dancing bubbles, rising and bursting, image emotion. It is the only water always clear and sparkling. Streams gather mud, springs dispel it. They come pure from the depths, and never suffer the earth to gather where they leap from ground. They are the brightest and the cleanest things in Nature. From all time the polluter of a spring has been held accursed.

One of the sources of the Thames was a real spring, rising from the earth in a meadow, until the level of the subterranean water was reduced.

These suddenly uprising springs are not common in our country, and need seeking. Our poets, who borrowed from the classics all their epithets for natural *fountains*, wrongly applied them to our modest springs welling gently from the bosom of the earth. The springs of old Greece and Italy gushed spouting from the rocks or flowed like the fountains of Tivoli in falling sheets over dripping shoots of stone. Even a Greek of to-day never speaks of a "spring," because he seldom sees one. "Fountain" is the word used for all waters flowing from the earth, and the difference of words corresponds to a difference of fact. The springs of his land *are* fountains, waters gushing from the rock or flowing from caverns and channels in the hills. The fountains of Greece flow down from above, and do not bubble up from below. These are the waters that tell their presence by sound, and have been the natural models of all the drinking fountains ever built,—jets that, spouting in a rainbow curve, hollow out basins below them, cut in the marble floor, cool cisterns ever running over, at which demi-gods watered their horses, and the white feet of the nymphs were seen dancing at sundown.

A tributary of the Severn, near Bisley, in the Cotswolds, bursts from a real fountain pouring from a hollow face of stone. But fountains in this sense are rare in England, though among the Welsh hills and the Yorkshire dales they may be seen springing full grown from the sides of the glens or "scarrs," and cutting basins and steps in marble or slate. But in the South the gentle springs take their place, silent, retiring, seldom found, except by chance, or by the local tradition which always attaches to the more important of our English natural wells. These it is the ambition of misdirected zeal to enclose in walls of stone, and to furnish with steps and conduits. If the old goddess Tan was once worshipped as the deity of the spring, it has usually undergone conversion by the early monks and changed its title to "St. Anne's Well," or been assigned to St. Catherine or some other of the holy sisterhood of saints.[1] But there are hundreds of tiny springs in Britain still left as Nature made them, and not yet settled in trust on any of the modern successors to the water rights of classic nymphs and Celtic goddesses. He who discovers for himself one of these springs will visit it each time he passes near. Some are in the woods, known only to the birds and beasts which live in them, and come daily to drink the pure, untainted waters. Wood springs are among the most beautiful of all, for they have a setting of tall timber, and their margins are never trampled by cattle, or the natural play of their waters disturbed to draw for the beasts of the farm. In the wood below Sinodun Hill there rises an everlasting spring. There may be seen how great an area of land it takes to make and keep one tiny spring. All the waters which gather in the millions of tons of chalk on Sinodun rise and flow out in the wood in the one pool, not larger than the circle of a wheel. It is always full, with the water throbbing up clear from the invisible vents below, and tiny white water-shells floating and falling in the basin, set round with liverwort and moss, and watering a bed of teazles in the wood below. Children drink from it, and pluck wild strawberries by its banks, and the pheasant and the fox come there to quench their thirst. An unexpected but not uncommon site of such springs is close to the margin of streams, which themselves are fed, not mainly by springs, but from the surface waters. [2] Wherever high ground slopes down to a stream, and ends in a rising bank at some distance from the river, there a true spring often rises, with an existence wholly apart from that of the river close by, into which its surplus of waters flows. Such springs have their special flora, their own "phenomena," and their own little set of effects on their liliput landscape. In the centre the waters well up, absolutely pure, and only discoloured when a more impatient earth-throb drives up a column of cloudy sand or earth. The spreading circles broaden outwards, and make their little marsh, planted with water-grass and forget-me-nots and blue bog-bean, and in the spring with butterburs. Outside, on the firmer but still moist soil the creeping jenny mats the ground; and the succulent grasses which attract the cattle to tread the marsh into a muddy paste. At the foot of the larger chalk downs the springs sometimes break out in different fashion, a modest imitation of classical fountains. The chalky soil breaks down, and from its sides the water often spouts in jets, as may be seen in Betterton glen, above Lockinge House, and in many other heads of the chalk brooks.

Springs of this kind are the natural outflowing of the water-bearing strata, where they lie upon others not pervious. But the upflowing springs are often fed by the accumulations of a great area of country, coming to the surface like water from the orifice of a syphon, and flowing permanently neither in greater nor less volume with constant force. If these cease to run the inference is that the old conditions are seriously disturbed. This has happened so frequently of late that local authorities would do well to schedule lists of the larger springs and request the owners or occupiers of the land to inform them from time to time whether there is a decrease in the flow. Stored water is almost as valuable as earth in a cycle of deficient rainfall, and the loss of any of our fountains and springs is a local misfortune not easily remedied.

[1] "Well deckings" are still common festivals in the North. Quite lately a Scotch loch was dragged with nets to catch a kelpie, and the bottom sowed with lime. The Church early forbade well worship.

[2] There is one such just above Marston Ferry, near Oxford, on the Cherwell, and two in a field below Ardington, near Lockinge.

BIRD MIGRATION DOWN THE THAMES

On September 16, 1896, after a period of very stormy wet weather, I saw a great migration of swallows down the Thames. It was a dark, dripping evening, and the thick osier bed on Chiswick Eyot was covered with wet leaf. Between five and six o'clock immense flights of swallows and martins suddenly appeared above the eyot, arriving, not in hundreds, but in thousands and tens of thousands. The air was thick with them, and their numbers increased from minute to minute. Part drifted above, in clouds, twisting round like soot in a smoke-wreath. Thousands kept sweeping just over the tops of the willows, skimming so thickly that the sky-line was almost blotted out for the height of from three to four feet. The quarter from which these armies of swallows came was at first undiscoverable. They might have been hatched, like gnats, from the river.

In time I discovered whence they came. They were literally "dropping from the sky." The flocks were travelling at a height at which they were quite invisible in the cloudy air, and from minute to minute they kept dropping down into sight, and so perpendicularly to the very surface of the river or of the eyot. One of these flocks dropped from the invisible regions to the lawn on the river bank on which I stood. Without exaggeration I may say that I saw them fall from the sky, for I was looking upwards, and saw them when first visible as descending specks. The plunge was perpendicular till within ten yards of the ground. Soon the high-flying crowds of birds drew down, and swept for a few minutes low over the willows, from end to end of the eyot, with a sound like the rush of water in a hydraulic pipe. Then by a common impulse the whole mass settled down from end to end of the island, upon the osiers. Those in the centre of the eyot were black with swallows—like the black blight on beans.

Next morning, at 6.30 a.m., every swallow was gone. In half an hour's watching not a bird was seen. Whether they went on during the night, or started at dawn, I know not. Probably the latter, for Gilbert White once found a heath covered with such a flock of migrating swallows, which did not leave till the sun dispelled the mists.

The migration routes of birds follow river valleys, when these are conveniently in line with the course they wish to take. There is far more food along a river than elsewhere, and this is a consideration, for most birds, in spite of the wonderful stories of thousand-mile flights, prefer to rest and feed when making long migrations, and also those short shifts of locality which temporary hard weather causes. A friend just back from Khartoum tells me that he saw the storks descending from vast heights to rest at night on the Nile sandbanks, and saw their departing flight early in the morning, these birds being in flocks of hundreds and thousands.

By watching the river carefully for many years I have noticed that it is a regular migration route for several species besides swallows. The first to begin the "trek" down the river are the early broods of water-wagtails, both yellow and pied. They turn up in small flocks so early in the summer that one might almost doubt if they could fly well enough to take care of themselves. On June 26th last summer nearly forty were flying about in the evening, and went across to roost on the eyot. Later numbers of blackbirds arrive, also moving down the river. Sand-martins, when beginning the migration, travel down the Thames in small flocks, and sleep each night in different osier beds. How many stages they make when "going easy" down the river no one knows. But I have seen the flocks come along just before dusk, straight down stream, and then dropping into an osier bed.

In the second week of September there is usually an immense migration of house-martins and swallows down the river. I have already described what I once saw on a migration night on Chiswick Eyot. Sometimes they go on past London, and find themselves near Thames mouth with no osier beds or shelter of any kind. Then they settle on ships. I was told that one morning the craft lying in Hole Haven off Canvey Island were covered with swallows, all too numb to move, but that when the sun came out the greater number flew away towards the sea. The same thing happened on the windmill at Cley, in Norfolk, a famous starting and alighting place for birds. Moorhens evidently migrate up or down the river in spring and autumn, and occasionally dabchicks; otherwise their sudden appearance and disappearance on the eyot could not be accounted for. Snipe follow the Thames up the valley. Formerly Chiswick Eyot was their first alighting place when east winds were blowing, after the fatigue of crossing London; and persons still living used to go out and shoot them. A friend of mine, whose family has resided in Chiswick for several generations, used to go down the outside of the eyot and kill snipe, and also kill teal and duck in the stream which runs from Chiswick House into the river. Another friend broke a young pointer to partridges on the market garden between Barnes Bridge and Chiswick.

Probably a number of the warblers also use the river as a migration road, though I only notice them in spring. But as I am never here in early September possibly many pass without being noticed. Also they are silent in autumn, whereas in spring they sing, a little, but enough to show that they are there.

Among the birds of this kind which pass up the river, but of which only a few pairs stay to breed on the eyot, are whitethroats, blackcaps, chiff-chaffs, and, I believe, nightingales. One beautiful early morning in spring I could not believe my ears, but I heard a nightingale in a bush by the side of the garden overhanging the river. It sang for about an hour, "practising" as nightingales do. Another person in a house near also heard it, and was equally astonished. It probably passed on, for next day it was inaudible.

In hard weather a migration of a different kind takes place down the river towards the sea. These birds are recruited from the ranks of the birds that stay, with some foreign winter visitors also. They pass down the river feeding on the mud and among the stones at ebb tide. Among those I have seen are flocks of starlings and scattered birds, mainly redwings, thrushes, blackbirds, and occasionally robins. Sandpipers also migrate up the Thames in spring, and down it in autumn.

WITTENHAM WOOD

In Wittenham Wood, which in our time was not spoiled, from a naturalist's point of view, by too much trapping or shooting the enemies of game, though there was plenty of wild game in it, the balance of nature was quite undisturbed. Of course we never shot a hawk or an owl, and I think the most important item of vermin killed was two cats, which were hung up as an awful instance of what we could do if we liked.

[Illustration: OTTERS. *From a photograph by J. S. Bond.*]

[Illustration: WATERHEN ON HER NEST. *From a photograph by R. B. Lodge.*]

In such large isolated woods, the wild life of the ordinary countryside exists under conditions somewhat differing from those found even in estates where the natural cover of woodland is broken up into copses and plantations. Birds and beasts, and even vegetation, are found in an intermediate stage between the wholly artificial life on cultivated land and the natural life in true forest districts like the New Forest or Exmoor. Most of these woods are cut bare, so far as the underwood extends, once in every seven years. But the cutting is always limited to a seventh of the wood. This leaves the ground covered with seven stages of growth, the large trees remaining unfelled. With the exception of this annual disturbance of a seventh of the area, and a few days' hunting and shooting, limited by the difficulty of beating such extensive tracts of cover, the wood remains undisturbed for the twelve months, and all wild animals are naturally tempted to make it a permanent home.

As I have said, the wood stands on the banks of the Thames, below the old fortress of Sinodun Hill, and opposite to the junction of the River Thame. All the British land carnivora except the martin cat and the wild cat are found in it. The writer recently saw the skin of a cat which had reverted to the exact size, colouring, and length of fur of the wild species, killed in the well-known Bagley Wood, an area of similar character, but of much greater extent, at a few miles distance in the direction of Oxford. A polecat was domiciled in Wittenham Wood as lately as August, 1898. Though this animal is reported to be very scarce in many counties, there is little doubt that in such woods it is far commoner than is generally believed. Being mainly a night-hunting animal it escapes notice. But in the quiet of the wood it lays aside its caution, and hunts boldly in the daytime. The cries of a young pheasant in distress, running through some thick bramble patches and clumps of hazel, suggested that some carnivorous animal was near, and on stepping into the thicket a large polecat was seen galloping through the brushwood. Its great size showed that it was a male, and the colour of its fur was to all appearance not the rich brown common to the polecat and the polecat cross in the ferret, but a glossy black. This, according to Mr. W.E. de Winton, perhaps the best authority on the British *mustelidae*, is the normal tint of the male polecat's fur in summer. "By the 1st of June," he writes, "the fur is entirely changed in both sexes. The female, or 'Jill,' changes her entire coat directly she has young; at the end of April or the beginning of May. The male, or 'Hob,' changes his more leisurely throughout the month of May. He is then known locally as the black ferret, and has a beautiful purplish black coat. As in all *mustelidae* the male is half as big again as the female." Stoats and weasels are of course attracted to the woods, where, abandoning their habit of methodical hedgerow hunting, they range at large, killing the rabbits in the open wood, and hunting them through the different squares into which the ground is divided with as much perseverance as a hound. They may be seen engaged in this occupation, during which they show little or no fear of man. They will stop when crossing a ride to pick up the scent of the hunted rabbit, and after following it into the next square, run back to have another look at the man they noticed as they went by, with an impudence peculiar to their race. The foxes have selected one of the prettiest tracts of the wood for their breeding-earth. It is dug in a gentle hollow, and at a height of some forty feet above the Thames. From it the cubs have beaten a regular path to the riverside, where they amuse themselves by catching frogs and young water-voles. The parent foxes do not, as a rule, kill much game in the wood itself, except when the cubs are young. They leave it early in the evening and prowl round the outsides, over the hill, and round the Celtic camp above, and beat the river-bank for a great distance up and down stream, catching water-hens and rats. At sunrise they return to the wood, and, as a rule, go to earth. The cubs, on the other hand, never leave it until disturbed by the hounds cub-hunting in September. Otters, which travel up and down the river, and occasionally lie in the osier-bed which joins the wood, complete the list of predatory quadrupeds which haunt it. With the exception of the first, the wild cat, and the last, the otter, they constitute its normal population, and as long as the stock of rabbits and hares is maintained, they may remain there as long as the wood lasts.

Numerically, the rabbits are more than equal to the total of other species, whether bird or beast.[1] In dry seasons, they swarm in the lighter tracts of the wood, and burrow in every part of it. These wood-rabbits differ in their way of life from those in the open warren outside. Their burrows are less intricate, and not massed together in numbers as in the open. On the other hand, the whole rabbit population of the one hundred acres seems to keep in touch, and occasionally moves in large bodies from one part of the area to another. During one spring and early summer the first broods of young rabbits burrowed tunnels under the wire-netting which encircled the boundary for many hundred yards, and went into a large field of barley adjoining. This they half destroyed. By the middle of August it was found that, instead of the barley being full of rabbits, it was deserted. They had all returned to the wood, and were in their turn bringing up young families. One colony deserted the wood altogether, and formed a separate warren some hundreds of yards away on a steep hillside. On the eastern boundary the river is a complete check to their migration. Except in the great frosts, when the Thames is frozen, no rabbit ever troubles to cross it. Hares do so frequently when coursed, and occasionally when under no pressure of danger. After harvest, when the last barley-fields are cut, the wood is full of hares. They resort to it from all quarters for shelter, and do not emerge in any number until after the fall of the leaf. During the months of August, September, and October these hares, which during the spring and winter lie out in the most open parts of the hills above, lead the life of woodland animals. In place of lying still in a form throughout the day, they move and feed. At all hours they may be heard fidgeting about in the underwood and "creeping" in the regularly used paths in the thick cover. When disturbed they never go at speed, but, confident in the shelter of the wood, hop and canter in circles, without leaving cover. In the evening they come out into the rides, and thence travel out into the clover layers, returning, like the foxes, early in the morning. A badger was found dead in the wood the first year I rented it. This I much regretted, for though it had probably been shot coming out of a cornfield next the wood, the badger is quite harmless, and most useful to the fox hunter, for he *cleans out the earths*. Mr. E. Dunn, late master of the Old Berkshire, tells me that they are of great service in this way, as they *dig* and enlarge the earths, and so prevent the taint of mange clinging to the sides if a mangy fox has lain in them.

[Illustration: DABCHICK. *From a photograph by R.B. Lodge.*]

[Illustration: BADGER. *From a photograph by J.S. Bond.*]

Lying between the river and the hills, this wood holds nearly every species of the larger woodland and riverine birds common to southern England. The hobby breeds there yearly. The wild pheasant, crow, sparrow-hawk, kestrel, magpie, jay, ringdove, brown owl, water-hen (on the river-bounded side), in summer the cuckoo and turtle-dove, are all found there, and, with the exception of the pigeons and kestrels, which seek their food at a distance during the day, they seldom leave the shelter of its trees. One other species frequents the more open parts of the cover in yearly greater numbers; this is the common grey partridge. The wood has an increasing attraction for them. They nest in it, fly to it at once for shelter when disturbed, lie in the thick copses during the heat of the day, and roost there at night. Several covies may be seen on the wing in a few minutes if the stubbles outside are disturbed in the evening, flying to the wood. There they alight, and run like pheasants, refusing to rise if followed. It is said that in the most thickly planted parts of Hampshire the partridge is becoming a woodland bird, like the ruffed grouse of North America. All that it needs to learn is how to perch in a tree, an art which the red-legged partridge possesses. The birds, unlike the foxes, hares, and rabbits, avoid the centre of the wood. Only the owls and wood-pigeons haunt the interior. All the other species live upon the edge. They dislike the darkness, and draw towards the sun. The jays keep mainly to one corner by the river. The sparrow-hawks have also their favourite corner. The wild pheasants lead a life in curious contrast to that of the tame birds in the preserves. Like their ancestors in China and the Caucasus, they prefer the osier-beds and reeds by the river to the higher and drier ground. But in common with all the other birds of the wood, with the exception of the brown owls, they move round the wood daily, *following the sun*. In the early morning they are on the eastern margin to meet the sunrise. At noon they move round to the south, and in the evening are on the stubbles to the west. Where the pheasants are there will the other birds be found, in an unconscious search for light. It is the shelter and safety of the big wood, and not the presence of crowded vegetation, that attracts them. They seek the wood, not from choice, but because it is a city of refuge.

[1] These observations were made some years ago. I believe it has been found necessary to kill down the rabbits since.

SPORT AT WITTENHAM

There is always some rivalry about shooting different woods on adjacent properties, and the villages near always take a certain interest in the results. Visiting our nearest riverside inn to order luncheon for our own shoot that week, I found about a dozen labourers in the front room, with a high settle before the fire to keep the draught out, sitting in a fine mixed odour of burning wood, beer, and pipes. Sport was the pervading topic, for a popular resident had been shooting his wood, and many of the men had been beating for him, and had their usual half-crown to spend. They were all talking over the day at the top of their voices; it had been a very good one. The wood is quite isolated and not more than forty acres. All round it is the property of one of the Oxford Colleges, which retains the sporting rights over about fifteen hundred acres. This is exercised by one of their senior fellows under some arrangement which works perfectly well so far as I can see. I asked our keeper, who always calls him "The Doctor," whether he was a medicine doctor or a doctor of divinity. He inclined to think he was the latter, as he belonged to college shooting. This way of putting it struck me as odd, but he was right. Any way, he looked a very pleasant figure in his long shooting coat and old-fashioned Bedford cords. There is also a college keeper, who is an institution in the village. The day's sport in "the Captain's wood" had been a success. Forty hares had been shot, or just one per acre, as well as a number of rabbits and wild pheasants. The hares were being sent round the village in very generous fashion, and a dozen lay on a bench in a back room.

Our own day was also a satisfactory one. Rabbits were unusually numerous, and many squares had to be beaten twice. The gross total of the two days was only something over three hundred head; but it was all wild game, and shot in very pretty surroundings. With the beaters were the keeper, who is also head woodman, and two assistant woodmen. These three men cut the whole of the hundred acres down in the course of seven years. Putting their lives at something over three score and ten, they will, as they began before they were twenty-one, have cut the wood down about eight times in the course of their existence. The beaters are entirely recruited from the staff of this very large and well-managed farm. They have beaten the woods so often that they know exactly what to do, when properly generalled. Our landlord was one of the guns, and his son, who does not shoot, but knows the wood thoroughly, kindly took command of the men, and kept things going at best pace through the day. Anything prettier than the entrance to the wood would be hard to find. A long meadow slopes steeply to the Thames, with an old church and the remains of a manor house at one end and the wood at the other. Below the house is a roaring weir, and opposite the abbey of Dorchester across the flats. Our little campaign gave an added interest to the scene. The bulk of the men were going round behind the hills to drive these "kopjes" into the wood. The guns and one or two ladies, and some small boys bearing burdens were walking up the middle ride. Below was the silver Thames in best autumn livery, for the leaf was not yet off the willows, though the reed-beds were bright russet. The sky was blue, the sun bright, and the sound of the weir came gaily up through the trees. All the wood-paths were bright with moss, the air still, and an endless shower of leaves from the oaks was falling over the whole hundred acres. There were just enough wild pheasants in the wood to make a variety in the rabbit-shooting. Hares were unexpectedly numerous, and we lined up on the side of the wood furthest from the river for a hare drive. The whole hillside is without a hedge. Watching the long slope it is a pretty and exciting sport to see the coveys of partridges, of which there are sometimes a number on the hill, rise, fly down and pitch again, and then rise once more and come fifty miles an hour over your head into the wood.

The hares are generally very wild, getting up while the folds of the ground are still between them and the beaters. As they seldom come straight into the wood it is amusing to guess which particular gun they will make for. Most of them slipped in at a safe distance, only to be picked up in the wood later. A few birds were shot, and the cover now held some forty partridges, though they are very wild in the low slop, and seldom leave more than one or two stragglers behind when the wood is beaten. The rabbit-shooting in the cover is difficult unless firing at "creepers" from the cover in front is indulged in. The rides are often very narrow, and the rabbits cross like lightning. Shooting "creepers" is also highly dangerous if there are many guns, or if the men are near. They do not seem to mind; indeed, I have known them shout out exhortations for us to fire, when only screened by a row of thistles. One thing I have learnt by shooting this big wood. The hares, and late in the season the rabbits, move at least one square ahead of the beaters. If a single gun is kept well forward, choosing his own place and taking turnabout with the others, the bag—if it is wished to kill down the ground game—will be considerably increased. One object when shooting this wood is to get the ground beaten quickly; if there are twenty squares to be beaten, and five minutes are wasted at each, it means a loss of one hour forty minutes. The guns consequently go best pace to their places forward after each beat. What with running at a jog-trot down the rides, shooting hard when in place, and then getting on quickly to the next stand, often along spongy or clayey rides on a nice, warm, moist November day, this is by no means the armchair work which people are fond of calling wood shooting. The variety of scenery in the wood added much to the charm. Sometimes we were in the narrow rides covered with short turf and almost arched over by the tall hazels; sometimes we were in low slop or walking through last year's cuttings, shooting at impossible rabbits. There we had an occasional rise of those most difficult of all birds to kill, partridge in cover, killing both French and English birds; or a cock pheasant would rise and hustle forward, an agreement having been made to leave these till properly beaten up later in the day. Two very pretty corners were perhaps the most enjoyable parts of the sport. By the river was a flat reed- and rush-covered corner, with a ring of oaks round, the Thames at the bottom, and some tall chestnut-trees on the outside. As the men advanced we had a regular rise of wild pheasants, rocketing up from the reeds in every direction high over the oaks and chestnuts. A fox helped the fun by trotting up and down in the reeds uncertain which way to go, and flushing the birds as he did so. Then the rushes were walked out and the rabbits sent darting in every direction. After this we hardly found a bird or rabbit in that corner during the season.

That year the wood gave constant sport, far better than in the later years. There were three times as many rabbits, as well as hares and pheasants.

One day in January we shot it during a fall of fine, dry snow. As the day went on the ground grew white, and our coats whiter. At luncheon the men were quite prepared for the emergency, or rather had prepared for it the day before when the frost began. They had a bonfire of brambles a dozen feet high, and faggots ready as seats, one set for us on one side of the fire, another for themselves on the other. The roaring blaze of the fire warmed us through and through, and by the end of luncheon our coats, which had been powdered with snow, were grey with wood ash descending. During this day a fox hung round us during the whole shoot. I think he must have been picking up and burying or hiding wounded rabbits, for every now and then he would come out into the ride, carefully smell the various places where rabbits had crossed, and then, selecting one, would go off like a retriever into the cover.

SPORT AT WITTENHAM (*continued*)

A month later Mr. Harcourt was shooting his woods at Nuneham. There are more than four hundred acres of woods round this most beautiful park, all of them giving ideal English estate scenery. The oaks of the park are like those at Richmond, but there is not much fern except in the covers. Nuneham is the best natural pheasant preserve in the Thames Valley, except Wytham, Lord Abingdon's place, above Oxford. The woods lie roughly in a ring round the park, in which the pheasants sun themselves. Outside these woods are arable fields with quantities of feed, and all along the front lies the river, which the pheasants do not often cross. The most striking sport at Nuneham is the driving of the island by the lock cottage. Every one who has been at Oxford has rowed down to have tea under the lovely hanging woods by the old lock. Few see it later in the year when the island opposite is covered with masses of silver-white clematis and thousands of red berries of the wild rose and thorn. In the late autumn mornings, when the mists are floating among the tall trees on the hill and the sunbeams just striking down through the vapours as they top the wood from the east, it is one of the prettiest sights on the Thames. In November or early December, when the woods are shot, numbers of pheasants are always found on the island. It holds a pool, in which and on the river are usually a number of wild ducks. Shooting from the river itself is now forbidden, and these and the half-wild duck have multiplied. The beaters, in white smocks, all cross the old rustic bridge like a procession of white-robed monks, and drive this island, and wild ducks and pheasants come out high over the river, making for the top of the hill. The shooting is fast and difficult, and the scene as the guns fire from the stations all along the bank is most picturesque.

Shooting with a neighbour on some land adjoining Nuneham, my attention was drawn to the very elegant appearance of all the gates and rails adjacent to the road. As the ground was always beautifully farmed and in good order, the condition of the gates did not surprise me. There was, however, a story attached to their smartness. A seller of quack medicines had sent out advertisers with most objectionable little bills, which he had posted on every gate adjoining the roads. My entertainer, who was the occupier of the land, had brought an action against the medicine man for defacing his gates, which was only compromised by the delinquent undertaking to paint every gate. He demurred at first to painting the railings too, but in the end had to do this also.

The stalking-horse is still part of the sporting equipment of some old Thames-valley farmhouses, but not in this neighbourhood. Only one wet season fell to my lot, and then, though I often saw bodies of duck, I had no opportunity of getting near them. A neighbour anchored a punt under a hedge on the line which he believed the duck would take at dusk, and killed several. Hard frosts send large bodies of duck to the river; they come as soon as ever the large private lakes, like those at Blenheim, Wootton, and Eynsham are frozen, and lie in small flocks all along the river. Water-hens are so numerous on the river now, owing to their preservation by the Conservancy, that any small covers of osier near are full of them. They make extremely pretty old-fashioned shooting when beaten up by a spaniel from the sedge and osier cover. I once turned out a dozen water-hens, a brown owl, a woodcock, and a water-rail from one little withe patch. When shooting the wood we always had one or two water-hens in the bag, and sometimes a chance at a duck flying overhead from the river. Only once were there many woodcocks in the cover. There must have been at least five, and all were missed. At last, as we were finishing the beat, one of the guns, who was young and keen, went off after the last-missed cock along the river bank. As we were loading up the game at the wood gate we heard a single shot. Then he appeared in the ride with the cock. Both he and his excellent old spaniel received warm congratulations. For my own part I was never tired of by-days in the wood in my first season. The best sport was starting rabbits from under the rows of fresh-felled ash and hazel poles, which the woodman called drills. They are about five feet high and seven feet through. The rabbits get under them in numbers, and sit there all day. We had an old retriever who was an expert at finding them. The next process was for the gun to clamber on to the top and stand knee-deep on the springing faggots, while a woodman on each side poked the rabbit out with a pole. He might bolt any way, and was under the next drill in a trice, so the shooting was quick. I bagged twelve one afternoon in this cheerful manner. Another great ambition of our lives was to get the better of the hill partridges. There were plenty of them, but they always dived into the wood, and were lost for the day. Only once did we score off them. We drove about sixty from the hills into the wood. There they were seen running along the rides like guinea fowls, but by placing a gun at the corner of the wood, and beating towards him, we killed nine brace.

A FEBRUARY FOX HUNT

When the Yeomanry left the hunting field for South Africa, and "registered" horses were commandeered by Government, fox hunting in counties where it is not the main business of life might be supposed to languish. As a matter of fact, it did not; and if the fields were smaller than usual, and a good many familiar faces missing, the master very properly felt that as he had his pack and there were plenty of foxes, he might as well employ the one and hunt the other, and keep up the spirits of the county by good, sound sport and plenty of it. Masters who take this view, and there are very few who do not, are public benefactors and shining examples; for it is not only the men who hunt who benefit vastly by the change and exhilaration which hunting brings in its train. The whole countryside enjoys a wholesome tonic by the frequent visits of the hounds, and the well-equipped company with them. Nothing cheers up the village, or cures the influenza, or brings oblivion of war news, or puts every one into conceit with themselves, so quickly, or leaves such a glow of sound satisfaction behind it. It would be odd if it did not, considering the amount of time, money, and trouble spent before the pack trots up to the green before the old grey church at eleven on a February morning. Wittenham Wood lies on the very edge of the Old Berkshire country, and as the river blocks all one side of it is naturally not one of the favourite meets. But at the time of writing, early in February a meet was duly advertised, and punctually to time the hounds were there. Some people seem to think that modern fox-hunting is not so thorough as it was in the past. We know better, and without imitating Mr. Jack Spraggon, or reminding every one present of that "two thousand five hundred—twenty-five 'undred—pounds a year" which Lord Scamperdale did or did not spend on his pack, are very well aware of what our master and the servants and the hounds had done that morning. The meet is on the edge of his country, sixteen miles from his house, and he has ridden over all the way, rising before the sun has got through more than the outside layer of the mists. There is no special honour and glory awaiting him in return. The cover to be drawn is surrounded for miles by deep and holding land now soaked with rain. A run of any distinction is most improbable. On the other hand, there will be plenty of hunting of a certain kind, and the chance of seeing it, for the wood is overlooked by lofty hills. Therefore, though the meet is small, the neighbourhood as a body expect to see plenty of the hounds, and turn up expectant, the farmers on their cobs, the young ladies on ponies and in dog-carts, and all the village who can be spared for an hour on foot, while the small boys regard each other with rapturous grins, and practise "holloaing" to improve their lung-power when the fox breaks. When the hounds appear—they have come nearly as far from the kennels as the master has from home—they are covered with road mud from foot to head. The gritty splashes have changed all the white and tan to grey, and made the black badger-pied. While some roll on the grass and push themselves along sideways to get clean, and others attempt the impossible task of licking the mud off their legs and feet, the older hounds, who are less self-conscious, poke their heads into the hands and against the chests of their ready-made friends, the village children, who rush in while the master and the field and lookers-on are exchanging courtesies, and embrace all the pack whom they can reach. Meantime the "assets" for the day's sport, the material complement on which this present assembly must rely for its day's hunting, lie in the cover and its contents. A hundred acres of wood, in all stages of growth, from the high thickets which the woodmen were felling yesterday, to the teazle and stump-studded slope which they cut last year, with the deep river below and the swelling hills above, is the cover.

[Illustration: FOX AND CUB. *From photographs by Charles Reid.*]

What the master would like would be that it should hold but one fox, that that fox should get away over the hills and on to the downs beyond as quickly as possible, and that he should never come back, but be killed three parishes away. But no one believes in such luck; and the local lookers-on do not in the least desire it. They want to see "a day's hunting" in the wood, and a fox to every half-dozen hounds. As a fact there are five foxes, not one, in the wood; and, passing from the general to the particular, we may explain how they came there. The heavy rains of the end of January filled all the drains, in which many foxes lie, so full of water that they abandoned them in sheer disgust, and took to the warm lying of the wood. Among these was a most attractive vixen, whose society kept the rest from leaving when the weather improved; consequently, the wood seemed full of foxes, none of which were disposed to leave it. When the pack trotted up to the main ride, and the huntsman's ringing voice sent them crashing into the four-years' growth by the river, a brace were lying snug and dry in the old ash-stumps. One slipped into the river at once and quietly swam to the opposite bank, while the other crept all along the outside hedge and curled up in the corner waiting on events. The vixen slipped into a badger earth under an old oak and stayed there, and a couple more dog-foxes moved on into four acres of low slop, brambles, shoots, and blackthorns, where they were winded by half the pack, while the other half were running the first fox up the fence. The crash and music of the hounds re-echoed from the trees and the enfolding hills above, the shrieking of the jays as they flit protesting from tree to tree, the hearty ring of the huntsman's voice cheering his hounds — surely all this should send each fox flying out over the fields beyond! But a fox has no nerves. He keeps his head with the coolness of a Red Indian, and a "slimness" all his own. The first fox doubles back along his tracks, crosses the big ride, twenty yards lower, just as that part of the pack which is hunting him flings on up the fence, and waits again till he hears them break out where he first stopped. From outside, where the field are waiting on a knoll which gives a downward view into the rolling acres of the wood, the rest of the pack are seen forcing another fox upwards towards the hills. The sight is as pretty as our woods can show. Down below the red coats of the master and huntsman move up the rides, and the heads and sterns of the broad line of hounds, now all clean and bright after brushing through the wood, rise and fall, appear and vanish, as they leap over or thrust through the low slop and brambles. In front, where a goyle runs up to a hollow of the hill, the ground has been cleared of wood, and the forest of tall teazle-tops is full of goldfinches, flying from seed-head to seed-head, too tame to mind the noise or care for anything but their breakfast. Yet even they gather and fly before the approaching tumult. Hares come hurrying out, and dash over the smooth hillside; magpies rise, poise themselves, slue round, and dive backwards into the wood; and then circumspect, lopping easily and lightly along, a fox crosses through the teazles, and slips down to a drain in the hollow; and see! another fox behind him, along the same path, and on the same errand, for each trots up to a covered drain, looks at it, and finding it stopped, pauses a second to think, and takes his resolve. One slips back into the wood, the other canters to the fence, rising the hill, looks out, whisks his brush and is off—across the turf, over the fifty-acre field of growing wheat, and away to the back of the hills. Half the pack are running the first fox, who has slipped back to the river, but with the other half every one gets clear off, and does his best over the awful ground. The mud explodes like shells as the hoofs crush into it, but somehow every one is across and away, and on to the green road and a line of sainfoin much sooner than could be expected. The fox can be seen crossing the back of the hill, looking big and red, and full of running; but after twenty-five minutes over all sorts of ground, from medium bad to "downright cruel," for the soaking rains have made a very pudding even of the pasture, the fox is run into and killed close to the Thames. No one need be sorry for him, for he had lived by theft and violence for the past two years, and was duly eaten himself by his natural enemies. Then back to the wood again, where the rest of the pack had been whipped off their fox, and were waiting dolefully to begin again, by which time the other foxes, of which two elected to stay, had resolved that come what might, they would stick to the wood, of which they knew every inch by heart; and by keeping under the river bank, sneaking under layers of felled brushwood, dodging along drains, and other devices, postponed their fate for two hours, when one was "chopped" and one broke away and was run till dark. This is not the kind of thing that keeps hunting alive, but it is the kind of day which occurs in most ordinary counties in February, and at which no one greatly grumbles. But if a slow woodland day is unattractive, the man who hunts in a modest way from London and wishes to be sure of a run has no lack of choice. Try, for instance, a day on the South Downs, five miles from the sea, on the vast uplands and among the furze-covered bottoms behind Beachy Head, when the snow-clouds are rolling in from the Channel and dusting the summits of the downs with white. There is at least the certainty of foxes, and of a gallop over the highest and soundest land in the South, and even "February fill-dike" cannot make the going heavy.

EWELME—A HISTORICAL RELIC

At the head of one of the smaller Thames tributaries, a few miles from the river, lies Ewelme, the ancient Aquelma, so called from the springing waters which rise there. There are trout in the brook and excellent water-cresses higher up, which are cultivated scientifically. Also there was a political row in Gladstonian days over an appointment to the living. But the real interest of this exceptionally beautiful Thames-valley village is that it is a survival, almost unchanged, of a "model village" made in the time of the Plantagenets. As such it deserves a place in any history, even a "natural" history, which deals with the river.

The village lies at the foot of the Chiltern Hills, not far from Dorchester. The persons who made it a model village just before the Wars of the Roses were William de la Pole, the first Duke of Suffolk, and his Duchess, Alice, the grandchild of Geoffrey Chaucer. The Duke, as every one knows, was for years the leading spirit in England during the early part of the reign of Henry VI., whose marriage with Margaret of Anjou he arranged in the hope of putting an end to the disastrous war with France. His murder in mid-Channel—when his relentless enemies followed him out to sea, took him from the ship in which he was going into exile, and beheaded him on the thwarts of an open boat—was the forerunner of the most ghastly chapters of blood and vengeance in civil feud ever known in this country. But the grace and dignity of his home life in his palace at Ewelme, with his Duchess to help him, are less well known, though the evidences of it remain little altered at the present day.

[Illustration: EWELME POOL. *From a photograph by Taunt & Co.*]

Of course there was a village there long before the Duke of Suffolk became possessed of it. It was such a perfect site that if any place in the country round were inhabited, Ewelme would have been first choice. The flow of water is one of the most striking natural features and amenities of the place. It is a natural spring, coming out from the chalk of the Chilterns, and forming immediately a lovely natural pool, under high, tree-grown banks. This is still exactly as it was in the ancient days. No water company has robbed it, and besides "The King's Pool," which is the old name of the water, there are overflowing streams in every direction, now used in careful irrigation for the growth of watercress, one of the prettiest of all forms of minor farming. Fertile land, shelter from gales by the overhanging hill, great trees, and abundance of ever-flowing water, are the natural commodities of the place. It was of some importance very early, for it gave its name to a Hundred. This hundred contains among other places Chalgrove, where Hampden received his death-wound. Ewelme belonged to the Chaucer family. The last male heir was Thomas, son of Geoffrey Chaucer the poet, who left an only daughter Alice, destined to become the greatest lady of her time. She married first the celebrated Earl of Salisbury, who was killed by a cannon-shot while inspecting the defences of Orleans during the siege which Joan of Arc raised. William de la Pole, then Earl of Suffolk, was appointed commander of the English forces in the Earl of Salisbury's place, and not only succeeded to his office, but also married his Countess, who now became Countess of Suffolk. It was long before either the Earl or his Countess could revisit Ewelme, where the Earl must have had some property before his marriage, for his elder brother, Earl Michael, was buried at the public expense in the church of Ewelme after his death at Agincourt. For seventeen years the Earl never left the war in France; but when Henry VI. was grown up he arranged the marriage with Margaret of Anjou, and did his best to promote peace. At this time Suffolk was the most powerful subject in the kingdom. He was made a Marquis, and finally a Duke, and his Duchess was granted the livery of the Garter. In 1424 they built a palace at Ewelme, and in due course rebuilt the church, founded a "hospital for thirteen poor men and two priests," and added to this a school. Palace, church, hospital, and school were all of the same period of architecture, and that the very best of its kind. Thus in the fifteenth century Ewelme was eminently a "one man" place, like most of the model villages of to-day. The palace was moated, and used as a prison as late as the Civil War. Margaret of Anjou was kept there in a kind of honourable confinement for a short time, for long after the Duke's murder the Duchess was in favour once more, in the triumph of the Yorkists, and Margaret, who had been her Queen and patroness, was given to her keeping as a prisoner both in her palace and later at Wallingford Castle. Henry VIII. spent his third honeymoon there, with Jane Seymour, and Prince Rupert lived in it during the Civil War. Later, only the banqueting hall remained, which was converted into a manor house.

But if the palace is gone, the church remains as evidence of the magnificence of the Duke's ideas on the subject of a village place of worship. He seems to have shared the apprehension felt by the Duke in Disraeli's novel "Tancred," that he might be accused of "under-building his position." In design it is very like another large church at Wingfield in Suffolk, where his hereditary possessions lay, and where he was buried after his murder, his body having been given to his widow. The same architect possibly supervised both, but of the two Ewelme Church is the finer. The interior is especially splendid, for in it are the tombs of the Chaucers, and the magnificent sepulchre of the Duchess herself, on which her emaciated figure lies wrapped in her shroud. This tomb of the Duchess Alice is one of the finest monuments of the kind in England. The other relic of the prosperity of Ewelme under the De la Poles is the hospital and school they founded. "God's House" is the name now given to it, and it is kept in good repair and used as an almshouse. The inner court is surrounded by cloisters, and the whole is in exactly the same condition as when it was built. The higher parts, constructed of brick, were the quarters of the priest and schoolmaster. The ruin and subsequent murder of the Duke, who adorned and beautified this model village in the early fifteenth century, took place in 1450. Nearly all France was lost, and in the hopes of conciliating the enemy, Maine and Anjou were given up by Suffolk's advice. He was accused of "selling" the provinces, and a number of vague but damaging charges were drawn up against him on evidence which would not be listened to now in any court or Parliament, except perhaps in a French State trial. Suffolk drew up a petition to the king, which shows among other things the drain which the French wars made on the lives and fortunes of the English nobles. After referring to the "odious and horrible language that runneth through the land almost in every common mouth, sounding to my highest charge and most heaviest slander," he reminded the King that his father had died in the siege of Harfleur, and his eldest brother at Agincourt; that two other brothers were killed at the battle of Jargeau, where he himself had been taken prisoner and had to pay £20,000 ransom; that while his fourth brother was hostage for him he died in the enemy's hands; and that he had borne arms for the King's father and himself "thirty-four winters," and had "abided in the war in France seventeen years without ever seeing this land." The King's favour secured that he should be banished instead of losing his head, for a State trial was never anything better than a judicial murder. The following is the letter written by an eye-witness to Sir John Paston, describing what then happened: "In the sight of all his men he was drawn out of the great ship into the boat, and there was an axe and a stock. And one of the lewdest men of the ship bade him lay down his head and he should be fairly ferd (dealt) with, and die on a sword. And he took a rusty sword and smote off his head with half-a-dozen strokes, and took away his gown of russet and his doublet of velvet mailed, and laid his body on the sands of Dover; and some say his head was set on a pole by it, and his men sit on the land by great circumstance and pray." The writer says, "I have so washed this bill with sorrowful tears that uneths ye shall not read it." The Countess survived his fall and lived to be great and powerful once more. Her son became the brother-in-law of sovereigns, and her grandchildren were princes and princesses.

EEL-TRAPS

Fish and flour go together as bye-products of nearly all our large rivers. The combination comes about thus: Wherever there is a water-mill, a mill cut is made to take the water to it. The larger the river, the bigger and deeper the mill cut and dam, unless the mill is built across an arm of the stream itself. This mill-dam, as every trout-fisher knows, holds the biggest fish, and where there are no trout, or few trout, it will be full of big fish, while in the pool below there are perhaps as many more. Of all the food fishes of our rivers the eel is really far the most important. He flourishes everywhere, in the smallest pools and brooks as well as in the largest rivers, and grows up to a weight of 9 lb. or 10 lb., and sometimes, though rarely, more. His price indicates his worth, and never falls below 10d. per lb. Consequently he is valuable as well as plentiful, and the millers know this well. On nearly all rivers the millers have eel-traps, some of the ancient sort being "bucks," made of withes, and worked by expensive, old-fashioned machinery like the mill gear. Another and most paying dodge of the machine-made order is worked in the mill itself, and makes an annexe to the mill-wheel.

I once spent an agreeable hour watching the making of barley meal and the catching of eels, literally side by side. It was sufficiently good fun to make me put my gun away for the afternoon, and give up a couple of hours' walk, with the chance of a duck, to watch the mill and eel-traps working.

They were both in a perfect old-world bye-end of the Thames Valley, in the meads at the back of the forgotten but perfect abbey of the third order at Dorchester, under the tall east window of which the River Thame was running bank full, fringed with giant poplars, from which the rooks were flying to look at their last year's nests in the abbey trees.

The mill was, as might be supposed, the Abbey Mill; but on driving up the lane I was surprised to see how good and large was the miller's house, a fine dwelling of red and grey brick; and what a length of frontage the old mill showed, built of wood, as most of them are, but with two sets of stones, and space for two wheels.

Only one was at work, and that was grinding barley-meal — meal from nasty, foreign barley full of dirt; but the miller had English barley-meal too, soft as velvet and sweet as a new-baked loaf. Stalactites of finest meal dust hung from every nail, peg, cobweb, and rope end on the walls, fine meal strewed the floor, coarse meal poured from the polished shoots, to which the sacks hung by bright steel hooks, and on both floors ancient grindstones stood like monuments of past work and energy, while below and beside all this dust and floury dryness roared the flooded waters of the dam and the beating floats of the wheel. "Have you any eels?" I asked. "Come and see," said the miller.

He stopped his wheel, unbolted the door, and we looked up the mill dam, two hundred yards long, straight as a line, embanked by double rows of ancient yews, the banks made and the trees planted by the monks five hundred years ago. Then we stepped into the wheel-house, where the water, all yellow and foaming, was pouring into two compartments set with iron gratings below, on which it rose and foamed. Seizing a long pole with prongs like walrus teeth, the miller felt below the water on the bars. "Here's one, anyway," he said, and by a dexterous haul scooped up a monster eel on to the floor. In a box which he hauled from the dam he had more, some of 5-lb. weight, which had come down with the flood — an easy and profitable fishery, for the eels can lie in the trap till he hauls them out, and sell well summer and winter. It pays as well as a poultry yard. Once he took a 9-lb. fish; 2-1/2 lb. to 4 lb. are common.

The eel-trap on the old Thames mill stream is imitated in other places where there is no mill. Thus at Mottisfont Abbey on the Test an old mill stream is used to work an hydraulic ram, and also to supply eels for the house; the water is diverted into the eel-trap, and the fish taken at any time. Another dodge for taking eels, which is not in the nature of what is called a "fixed engine," is the movable eel-trap or "grig wheel." It is like a crayfish basket, and is in fact the same thing, only rather larger. They can be obtained from that old river hand, Mr. Bambridge, at Eton, weighted, stoppered, and ready for use, for 7s. 6d. each, and unweighted for 5s. They are neat wicker-work tunnels, with the usual contrivance at the mouth to make the entrance of the eels agreeable and their exit impossible. The "sporting" side of these traps is that a good deal of judgment is needed to set them in the right places in a river. Many people think that eels like carrion and favour mud. Mr. Bambridge says his experience is different, and his "advice to those about to fish" with this kind of eel-trap is suggestive of new ideas about eels. He says that "for bait nothing can beat about a dozen and a-half of small or medium live gudgeon, failing these large minnows, small dace, roach, loach, &c., though in some streams about a dozen good bright large lob worms, threaded on a copper wire and suspended inside, are very effective, and should always be given a trial. Offal I have tried but found useless, eels being a cleaner feeding fish than many are aware of; and feeding principally in gravelly, weedy parts, the basket should be well tucked up under a long flowing weed, as it is to these places they go for food, such as the ground fish, loach, miller's thumb, crayfish, shrimps, mussels, &c. When I worked a fishery near here, I made it a rule after setting the basket to well scratch the soil in front of the entrance with the boathook I used for lowering them, and firmly believe their curiosity was excited by the disturbed gravel. Choose water from four feet to six feet deep, and see basket lays flat. Every morning when picked up, lay them on the bank, pick out all weed and rubbish, and brush them over with a bass broom, keeping them out of water till setting again at dusk."

Eel-bucks, of which few perfect sets now remain, are the fixed engines so often seen on the Thames, and are a costly and rather striking contrivance, adding greatly to the picturesqueness of parts of the river. They are very ancient, and date from days when the "eel-run" was one of the annual events of river life. The eels went down in millions to the sea, and the elvers came up in such tens of millions that they made a black margin to the river on either side by the bank, where they swam because the current was there weakest. The large eels were taken, and are still taken, on their downward journey in autumn. It is then that the Thames fills, and at the first big rush of water the eels begin to descend to reach the mud and sands at the Thames mouth, where they spawn. They always travel by night, and it is then that the heavy eel-bucks are lowered. Often hundredweights are taken in a night, all of good size, one of the largest of which there is any record being one of 15 lb., taken in the Kennet near Newbury. In the "grig-wheels" they are taken as small as 3 oz. or 4 oz.; but in the bucks they rarely weigh less than 1 lb.

The darkest nights are the most favourable. Moonlight stops them, and they do not like still weather. The upward migration of eels goes on from February till May on the Thames, but the regular "eelfare" of the young grigs do not assume any great size till May, when as many as 1,800, about three inches long, were seen to pass a given point in one minute. So say the records. But who could have counted them so fast?

A few recent developments of the eel trade elsewhere show how valuable this may be. Quite lately the Danes discovered that the Lim-fiord and some other shallow Broads on the West Danish Coast were a huge preserve of eels. They began trawling there steadily, and have established a large and lucrative trade in them. On the Bann, in Ireland, eel catching is still done in a large way, and the fish shipped to London. But the most ancient and yet most modern of eel fisheries is on the Adriatic, at Comacchio, where lagoons 140 miles in circumference are stocked with eels, and eel breeding and exporting are carried out on a large scale. Even as early as the sixteenth century the Popes used to derive an income of £12,000 from this source.

SHEEP, PLAIN AND COLOURED

In the Thames Valley there are two very distinguished breeds of sheep—the Cotswolds at the head of the watershed, and the Oxford Downs, near Wallingford. Wallingford lamb is supposed to be the best in the market. There are also the Berkshire Downs sheep, but these are, I think, more obviously cross-bred, or else of the Hampshire breed. The Cotswold sheep are probably a very old breed. They are evidently the original of the woolly "baa-lamb" of the nursery, with long, fleecy wool. The Oxford Downs are a short-woolled sheep. One of the flocks of this breed has been improved by selection, mainly in regard to fecundity, to such an extent that I believe twins are the normal proportion among the lambs. The shepherds, as elsewhere on the large down farms, form a race apart.

They are not always on the best of terms with the ordinary farm labourers, I notice. "The shepherd be a working against I," is a complaint I sometimes hear. The real reason is that the shepherd thinks, above all things, of his flock, and of finding them *food*. The feud between the keeper of sheep and the raiser of crops dates from the days of Cain and Abel.

I heard lately from a gentleman who very frequently occupies the honourable position of judge or steward at the leading agricultural shows, that it is proposed that in future no sheep sent to shows are to be allowed to have their coats rouged, and the judges are in future to make their decisions uninfluenced by the beauties of cosmetics. This decision comes as a great blow to the skilled hands in the business of the "improver," who, by long experience and a nice knowledge of the weaknesses of judges, had brought the art of "making up" pedigree sheep of any particular breed to something very nearly approaching the ideal of perfection. Their wool was clipped so artistically as to resemble a bed of moss, and this being elegantly tinted with rouge or saffron, the sheep assumed the hue of the pink or primrose, according to taste and fancy. The reason for the demand which now requires that the champions of the flock shall be shown "plain" and not coloured is not too technical to appeal to the general public. Those who know the acute anxiety with which the exhibitors of prize animals, from fancy mice to shorthorns, watch them "coming on" as the hour for the show approaches, will treat tenderly, even if they cannot condone, the little weaknesses into which the uses of rouge and saffron led them. When a Southdown which ought to have a contour smooth and rounded as a pear still showed aggravating little pits and hollows where there ought to be none, nothing was easier than to postpone clipping those undesirable hollows till the moment before the show, or if there were bumps where there should be no bumps, to shave the wool down close over them. Left to Nature, the newly-clipped wool would show a different tint from the rest of the fleece; but the rouge or saffron then applied made all things even, to the eye, and the judges to find out whether the animals were "level" or not had to feel them all over. Feeling every six inches of some two hundred sheep's backs is very tiring work; so the judges have struck against rouge, and there is an end of it.

One night, some years ago, an extraordinary thing happened on both lines of downs by the Thames, near Reading, and also along the Chilterns. Most of the flocks over a very large area took a panic and burst from their folds, and next morning thousands of sheep were wandering all over the hills. I feel certain that there must have been an earthquake shock that night. Nothing else could have accounted for such a wide and general stampede. The last authenticated earthquake shock in the South Midlands took place hereabouts in 1775, and was noted at Lord Macclesfield's Castle of Shirbourne, where the water in the moat was seen to rise against the wall of one of the towers.[1]

Are our domestic sheep, except for their highly artificial development of wool, really very different from their wild ancestors, the active and flat-coated animals which still feed on the stony mountain-tops? The ways of sheep, not only in this country but abroad, show that a part at least of their wild nature is still strong in them; and if type photographs of all the representative domestic animals of our time, had been possible a few centuries ago, it may be that even in this country the shape of the animal would be found to have been far nearer to the sheep of St. Kilda and of the wild breeds than it is to-day.

In one of the old Cloth Halls of Norfolk are two fine reliefs in plaster, one showing the *Argo*, bringing the golden fleece, the other a flock of sheep of the day, with a saint in Bishop's mitre and robes preaching to them. The shepherd, in a smock, is spinning wool with a distaff; and the sheep feeding around him, though carefully modelled, are quite unlike any of the modern breeds. Many of the domestic sheep of hot countries are more slender and less woolly than the wild sheep of the mountains. The black-and-white Somali sheep, for instance, are as smooth as a pointer dog.

But it is in temperament and habits that the close connection between the wild and tame breeds is most clearly shown. The *excessive* domestication of the flocks of Southern England has killed all interest in them even among those who live in the country, and are keen and sympathetic observers of the ways of every other creature in the fields. The beauty of the lambs attracts attention, and the prettiness of the scene when they and their mothers are placed in some sheltered orchard among the wild daffodils and primroses, or in an early meadow by the brook, makes people wonder why they are so stupid when grown up. But the fact is that when not penned up by hurdles and moved from square to square over a whole farm, so that each inch of food may be devoured, each member of the flock can think for itself, and would, in less artificial surroundings, make for itself a creditable name for independence and intelligence. All sheep have retained this distinguishing habit of their ancestors, that they are by nature migratory, and share with nearly all migrant animals a capacity for thought and organisation, and a knowledge of localities. Wild sheep are migratory because they live by preference on the rocky and stony parts of hills just below the snow-line. This is why the tame sheep do so well on the moors of Scotland and mountains of Switzerland. But as the snow-line descends each winter far below their summer feeding haunts, wild sheep either migrate to the lower slopes of the mountains, or, like the deer of the Rockies, move off altogether to great distances. Every winter, for instance, the lower valleys of Yellowstone Park are filled with deer and antelope from the distant mountains. So the tame flocks of Greece, Thrace, Spain, and even Scotland are migratory. In Scotland their transport is modernised, and they travel regularly by steamer from the islands to winter in the Lowlands, and by train from the Highlands. Two years ago a flock of migratory sheep from Ayrshire came for early spring feeding to Hyde Park, and were there shorn, with their Highland collies looking on. In the "old countries" and the non-progressive East of Europe the migration of the flocks is on a vaster and far more romantic scale. In Spain there are some ten millions of migratory sheep, which every year travel as much as two hundred miles from the plains to the "delectable mountains," where the shepherds feed them till the snows descend. These sheep are known as *transhumanies* and their march, resting places, and behaviour are regulated by ancient and special laws and tribunals dating from the fourteenth century. At certain times no one is allowed to travel on the same route as the sheep, which have a right to graze on all open and common land on the way, and for which a road ninety yards wide must be left on all enclosed and private property. The shepherds lead the flocks, the sheep follow, and the flock is accompanied by mules carrying provisions, and large dogs which act as guards against the wolves. The Merino sheep travel four hundred miles to the mountains, and the total time spent on the migration there and back is fourteen weeks. In Thrace the migration of the flocks is to the northern ranges of Mount Rhodope. The sheep are said to be no less alert than the Pomak shepherds, obeying a signal to assemble at any moment given by the shepherd's horn. The dogs are ferocious in the extreme, as the enemies of sheep in these parts are more commonly men than wild beasts, and the gentle shepherd, who has, since the Russo-Turkish War, exchanged his long gun for a Winchester rifle, shoots at sight and asks no questions.

The more nearly domestic sheep can approach the life of the primitive stock, the more intelligent their way of life becomes. The cleverest sheep live on the hills, and the stupidest on the plains. In Wales, for instance, if a new tenant takes over the flock of an outgoing tenant, the latter is by law allowed a higher price if the flock is one which knows the boundaries and paths on the hills. On the plains of Argentina, as Mr. Hudson tells us, the lambs are born so stupid that they will run after a puff-ball rolling before the wind, mistaking it for their mother.

[1] This was a tremor of the great earthquake at Lisbon.

SOME RESULTS OF WILD-BIRD PROTECTION

Among the happiest results of the modern feeling about birds is the conversion of the whole of the Thames above the tideway into a "protected area." This was not done by an order of the Secretary of State, who, by existing law, would have had to make orders for each bit of the river in different counties, and often, where it divides counties, would have been obliged to deal separately with each bank. The Thames Conservancy used their powers, and summarily put a stop to shooting on the river throughout their whole jurisdiction. The effect of this was not seen all at once; but little by little the waterfowl began to return, the kingfishers to increase, and all the birds along the banks grew tamer. Then the County Councils of

Oxfordshire, Buckinghamshire, and Berkshire forbade the killing of owls and kingfishers, and this practically made the river and its banks a sanctuary.

The water-hen are so numerous that at Nuneham Lock they run into the cottages, and at other locks the men complain they eat all their winter cabbages. As many as forty at a time have been counted on the meadows. Mr. Harcourt has also established a wild-duck colony on and about the island at Nuneham. The island has a pond in the centre, with sedges and ancient willows and tall trees round. There the really wild ducks join the home-bred ones in winter. Lower down, the scene on late summer days is almost like a poultry-yard, with waterfowl and wild pigeons substituted for ducks and chickens. Young water-hens of all sizes pipe and flutter in the reeds, and feed on the bank within a few feet of those rowing or fishing, and their only enemies are the cats, which, attracted by their numbers, leave the cottages for the river and stalk them, while the old water-hens in vain try to get their too tame young safe on to the water again.

Though kingfishers have increased fast they are less in evidence, being naturally shy after years of persecution. In summer they keep mainly at the back of the willows, away from the river, so long as the latter is crowded with boats.

It was not till November, 1899, that I saw the kingfishers at play, as I had long hoped to do, in such numbers as to make a real feature on the river. It was a brilliant, warm, sunny morning, such as sometimes comes in early winter, and I went down before breakfast to Clifton Bridge. There the shrill cry of the kingfishers was heard on all sides, and I counted seven, chasing each other over the water, darting in swift flight round and round the pool, and perching on the cam-shedding in a row to rest. Presently two flew up and hovered together, like kestrels, over the stream. One suddenly plunged, came up with a fish, and flying to the other, which was still hovering, put the fish into its beak. After this pretty gift and acceptance both flew to the willows, where, let us hope, they shared their breakfast.

In a row down the river extending over ten miles I saw more than twenty kingfishers, most of them flying out, as is their custom, on the side of the willows and osiers averse from the river, but some being quite content to remain on their perches from which they fish, while the boat slipped down in midstream. As they sit absolutely motionless, and the reddish breast, and not the brilliant back, is turned to the water, it needs quick eyes to see these watchers by the stream.

The total prohibition of shooting on the water or banks is also producing the usual effect on the other birds and beasts. They are rapidly becoming tame, and the oarsman has the singular pleasure of floating down among all kinds of birds which do not regard him as an enemy. Young swallows sit fearlessly on the dead willow boughs to be fed by their parents; the reed-buntings and sedge-warblers scarcely move when the oar dips near the sedge on which they sit; wood-pigeons sit on the margin and drink where the pebble-banks or cattle-ways touch the water; and the water-rats will scarcely stop their business of peeling rushes to eat the pith, even if a boatload of children passes by.

[Illustration: NIGHTJAR AND YOUNG ONE. *From a photograph by R.B. Lodge.*]

[Illustration: REED BUNTING. *From a photograph by R.B. Lodge.*]

The return of the birds, and especially of wild fowl, to the London river is the result partly of the same causes which have restored the fish to its waters; partly, also, of measures affecting a wider area, but carried out with far less physical difficulty. Their presence is evidence that the tidal Thames now yields them a stock of food so abundant as to tempt birds like the heron, the water-hen, and the kingfisher back to their old haunts. It shows, secondly, that the by-laws for the protection of birds passed by the counties of London, Surrey, and Middlesex, and by the Thames Conservancy (which was the pioneer in this direction by forbidding shooting on the river), are so far effective that the stock is rapidly increasing; and, lastly, that the birds are preserved and left in peace to a great extent on the London river itself. The following are the most marked instances of this return of river fowl which have come under the writer's notice; but in every case there have been preliminary advances on the part of the birds, which show that what is now recorded is only one step further in the general tendency to resume their old habits, or even to go beyond their former limits of place and time in resorting to the river. The herons from Richmond Park have extended their usual nightly fishing ground, which formerly ended at Kew Bridge, four miles further down the river, almost to Hammersmith Bridge, and in place of coming late at night, under cover of darkness, have made a practice of flying down at dusk, and pitching on the edge of Chiswick Eyot.[1] Their regular appearance led to various inquiries as to the nature of the "big birds like geese" which flew down the river and made a noise in the evening, questions which were answered, in one case, by the appearance of one of the birds as it swung round in the air opposite a terrace of houses, and dropped in the stream to fish, not twenty yards from the road. As the heron is naturally among the shyest of all waterside birds, and seeks solitude above all things, these visits show that the quantity of fish in the lower river must be great, and also that the London herons, now never shot at, are losing their inbred dislike of houses and humanity. Their footprints have been found on the mud opposite a creek in Hammersmith, round which is one of the most crowded quarters of the poorer folk of West London. The birds had been fishing within ten yards of the houses, which at this point are largely inhabited by organ-grinders and vendors of ice-creams, callings which do not promote quiet and solitude in the immediate neighbourhood. In the evening and early morning a few wild ducks accompany the herons as low as the reach above Hammersmith Bridge, and single ducks have been seen even at midday flying overhead. At sunrise one Midsummer Day I saw a sheldrake (probably an escaped bird) flying down the river, looking very splendid in its black, white, and red plumage, in the bright light of the morning. It haunted the reach for some days, and was not shot. Among other visitors to this part of the river and its island during spring were a curlew, which fed for some time on the eyot during the early morning, and a pair of pheasants, one of which, an old-fashioned English cock bird, was subsequently captured unhurt. A flock of sandpipers remained there for some weeks, and during the summer numbers of sedge-warblers have nested on and around the eyot; the cuckoo has been a regular visitor to the osier-bed in the early morning, probably with a view to laying its eggs in the sedge-warblers' nests. As a set-off to these early visits of the cuckoo, a nightjar has hunted round the islet for moths, both at dusk and during the night, when its note may often be heard. This is a fairly long list of interesting birds revisiting a portion of the river which the London boundary crosses. At a distance of less than half a mile, on some ornamental water near the river, an even more unexpected increase of the bird population has been noted. A pair of kingfishers nested and reared their brood in an old gravel-pit, while several nests of young dabchicks hatched by the pool.[2] There also during the spring a pair of tufted ducks appeared, and remained for some days before going on their journey to their breeding haunts. One lamentable event in the bird life of the Thames deserves mention. A pair of swans ventured to nest within a few hundred feet of the London boundary. The hen, a very shy young bird, laid three eggs on Chiswick Eyot, and the pair, being supplied with material, diligently built up their nest day by day until it was above the tide level. They sat for five weeks, the cock bird keeping anxious guard day and night, while the hen would probably have died of starvation unless fed by kindly neighbours, for the river affords very little food for a swan, and this required far longer time to find than the bird was willing to spare from her nest. This was then robbed in the night, and the cock bird maltreated in defending it. The return of fish and fowl to the London Thames shows by the best of tests that the efforts of the Thames Conservancy to preserve the amenities of the river, of the Sewage Committee of the County Council to maintain its purity, or rather to render it less impure at its mouth, and of the adjacent County Authorities to protect bird life, are all yielding good results, and justify the courage with which such an apparently hopeless task was undertaken. To the Conservancy I would offer one or two suggestions, which County Councillors might also consider. The river is the only large *natural* feature still left in the area of London and Greater London. Now that it contains water in place of sewage, there is a guarantee that its main element as a natural amenity in a great city will be maintained, and as it becomes purer, so will the facilities which it offers for boating, fishing, and bathing increase. But it should not be *embanked* beyond the present limit at Putney. Stone walls are not a thing of beauty, and a natural river-bank is. At present, from Putney to Richmond the greater part of the Thames flows between natural boundaries. If these can be maintained, the growth of willows, sedge, hemlock, reeds, water ranunculus, and many other fine and luxuriant plants affords insect food for the fish and shelter for the birds, besides giving to the river its natural floral border. If this is replaced by stone banks the birds and the fish will move elsewhere.

[1] Mr. J.E. Vincent tells me that in 1902 the herons were heard as far down the river as Chelsea.

[2] In the beautiful grounds of Chiswick House, where the present occupier, Dr. T. Tuke, carefully preserves all wild birds.

OSIERS AND WATER-CRESS

Osiers, the shoots of which are cut yearly for making baskets, crates, lobster-pots, and eel-traps are a form of crop of which not nearly as much is made in the Thames Valley as their profitable return warrants. Properly managed they nearly always pay well, and, in addition, they are very ornamental, and for the whole of the summer, autumn, and winter are one of the very best forms of covert for game. They are commonly seen near rivers, especially in parts where the ground is flooded in winter. But osiers may be grown anywhere on good ground, and are a rapid and paying crop, giving very little trouble, though they need some attention even on the banks of tidal rivers. It is estimated that in the whole of Great Britain there are only between 7,000 and 8,000 acres of osier beds, but these average three tons of rods per acre, and the value of the crop when harvested is often at least £15 per acre gross return. As fruit cultivation is immensely increasing in England, there is a corresponding increase in the demand for baskets to put the fruit in. This is the main reason why osiers, unlike most farm crops, keep up their price. Immense quantities are now imported from Belgium, France, and Germany because our own crop is not nearly sufficient.[1] They do not require a wet soil or to be near water: all that the willow roots need is that the land shall be good and not too dry or sandy. Stagnant, boggy ground does not suit them at all, though they will grow well in light loam. Many species of osier are of most brilliant colouring in winter and early spring. In some the rods are golden yellow; in others the bark is almost scarlet with a bright polish, and the osier bed forms a brilliant object from December to February, just before the rods are cut. The kind of willow grown varies from the slender, tough withes used in making small baskets and eel-traps, to the large, fast-growing rods suited for making crates for heavy goods. The planter must find out for which kind there is the readiest market in the neighbourhood, and then get his land ready. It needs thorough clearing and trenching to the depth of from twenty to thirty inches. The young osiers should then be put in. These should be taken from a nursery in which they have been "schooled" for one year, as in that case they will produce a crop fit to cut one year earlier than if the cuttings have been put at once in the new osier-bed. The cuttings when transferred to the bed should be put in twelve inches apart in the rows, and these rows made at two feet distance from each other. They will need hoeing to keep the ground clear, which will cost £1 to £2 per acre for the first two years, and this should be done before the middle of June. When the osiers are well started they grow so densely that they kill out the weeds themselves. The rate of growth even on ordinary field-land is astonishing; they will add eighteen inches in a week. February and March are the months for planting, and March also sees the osier harvest when the time comes to cut them. In the fens the harvesting of the rods begins earlier, but this depends usually on the season, the object being to cut them before the sap begins to rise. Osiers particularly invite the attention of those who are desirous of planting coverts for game. They are a paying crop, and a quick crop, giving cover sooner and of better quality than almost any other form of underwood, and are also very ornamental. It is true that they are cut yearly, but this is not till the shooting season is over. Meantime there is no covert which pheasants like so much as osier-beds, especially if they are near water.

On Chiswick Eyot, which is entirely planted with osiers, there are standing at the time of writing six stacks of bundles set upright. Each stack contains about fifty bundles of the finest rods, nine feet high. Thus the eyot yields at least three hundred bundles. This osier-bed is cut quite early in the year, usually in January, and by February all the fresh rods are planted. Before being peeled the osiers are stood upright in water for a month, and some begin to bud again. This is to make the sap run up, I presume, by which means the bark comes off more readily. I believe that the Chiswick osiers, being of the largest size, are used for making crates, and that they are cut early because there is no need to peel them.

Water-cress growing is an increasing business in the Thames Valley, where the head of every little brook or river in the chalk is used for this purpose. This is good both for business in general and for the fish, for water-cress causes the accumulation of a vast quantity of fish food in various forms.

The artificial culture of water-cress is comparatively modern, and a remarkably pretty side-industry of the country.

Formerly, the cress gatherer was usually a gipsy, or "vagrom man," who wandered up to the springs and by the head waters of brooks at dawn, and took his cresses as the mushroom-gatherer takes mushrooms — by dint of early rising and trespass.

[Illustration: PEELING OSIERS. *From a photograph by Taunt & Co.*]

The places where water-cress grows naturally are usually singularly attractive. The plant grows best where springs actually bubble from the ground, either where the waters break out on the lower sides of the chalk downs, or in some limestone-begotten stream where springs rise, sometimes for a distance of one or two miles, bubbling and swelling in the very bed of the brook. There, among dead reeds and flags, the pale green cresses appear very early in the spring, for the water is always warmer which rises from the bosom of the earth. Trout and wild duck haunt the same spots, and one often sees, stuck on a board in the stream, a notice warning off the poor water-cress gatherer, who was supposed to poach the fish.

The happy-go-lucky cress gathering is now a thing of the past, and there are few rural industries more skilfully and profitably conducted. I knew a farmer who, having lost all his capital on a large farm on the downs, took as a last resource to growing the humble "creases" by the springs below. He has now made money once more, and been able to take and cultivate another farm nearly as large as that he worked before, while the area of his water-cress beds still grows.

Wherever a chalk stream, however small, breaks out of the hills, it is usual to let it to a water-cress grower. He widens the channels, and year by year every square foot of the upper waters is planted with cress. Each year, too, new and larger beds are added below, and the cresses creep down the stream. When they encroach on good spawning ground this is very bad for trout; but the beds are pretty enough, forming successive flats, on different levels, of vivid green.

The scene on the Water-cress Farm shows the complete metamorphosis undergone by what was once a swift running brook when once the new culture is taken in hand. When left to Nature, the little chalk stream might truly have said, in the words of the poem—

"I murmur under moon and stars
 In brambly wildernesses,
 I linger by my shingly bars,
 I loiter round my cresses."

Now all the brambles and shingle are gone, and the stream is condemned to "loiter round its cresses," and to do nothing else. The water must not be more than six inches deep, and it must not flow too fast. To secure these conditions little dams, some made of earth and some of boards, are built from side to side of the brook. The water thus appears to descend in a series of steps, each communicating with the next by earthen pipes, through which the water spouts. When a fresh bed of cresses is to be planted, which is done usually towards the end of summer, a sluice is opened, and only an inch or so of water left. On this cuttings from the cress are strewn, which soon take root, and make a bed fit for gathering by next spring.

From February to April the cresses are at their best. Their flavour is good, their leaves crisp, and they come at a time when no outdoor salad can be grown. As the beds are set close to the fresh springs, they are seldom frozen. Hence, in very hard weather all the birds flock to the cress-beds, where they find running water and a certain quantity of food. If the beds do freeze, the cress is destroyed, and the loss is very serious.

Gathering cresses is a very pleasant job in summer, but in early spring one of the most cheerless occupations conceivable short of gathering Iceland moss. The men wear waterproof boots, reaching up the thighs, and thick stockings inside these. But the water is icy cold. The cress plants are then not tall, as they are later, but short and bushy. They need careful picking, too, in order not to injure the second crop. Then the cold and dripping cresses have to be trimmed, tied into bundles, and packed. When "dressed" they are laid in strong, flat hampers, called "flats," the lids of which are squeezed down tight on to them. The edges are then cut neatly with a sharp knife, and the baskets placed in running water, until the carts are ready to drive them to the station. Not London only but the great towns of the North consume the cress grown in the South of England. A great part of that grown in the springs which break out under the Berkshire Downs goes to Manchester.

One basket holds about two hundred large bunches. From each of these a dozen of the small bunches retailed at a penny each can be made; and every square rod of the cress-bed yields two baskets at a cutting.

In one of the East London suburbs, near to the reservoirs of a water company, it has been found worth while to create an artificial spring, by making an arrangement with the waterworks for a constant supply. This flows from a stand-pipe and irrigates the cress-beds, which produce good cresses, though not of such fine flavour as those grown in natural spring water and upon a chalk soil.

[1] Fishermen in the Isle of Wight send all the way to the Midlands to get the little scarlet withes required for making lobster-pots.

FOG AND DEW PONDS

The cycle of dry seasons seems to be indefinitely prolonged. During the period, now lasting since 1893, in which we have had practically no wet summers, and many very hot ones, a very curious phenomenon has been remarked upon the high and dry chalk downs.

The dew ponds, so called because they are believed to be fed by dew and vapours, and not by rain, have kept their water, while the deeper ponds in the valleys have often failed. The shepherds on the downs are careful observers of these ponds, because if they run dry they have to take their sheep to a distance or draw water for them from very deep wells. They maintain that there are on the downs some dew ponds which have never been known to run dry. Others which do run dry do so because the bottom is injured by driving sheep into them and so perforating the bed when the water is shallow, and not from the failure of the invisible means of supply. There seem to be two sources whence these ponds draw water, the dew and the fogs. Summer fogs are very common at night on the high downs, though people who go to bed and get up at normal hours do not know of them. These fogs are so wet that a man riding up on to the hills at 4 a.m. may find his clothes wringing wet, and every tree dripping water, just as during the first week of last November in London many trees distilled pools of water from the fog, as if it had been pouring with rain. Such was the case on July 4th, 1901. The fogs will draw up the hollows towards the ponds, and hang densely round them. Fog and dew may or may not come together; but generally there is a heavy dew deposit on the grass when a fog lies on the hills. After such fogs, though rain may not have fallen for a month, and there is no water channel or spring near the dew pond, the water in it rises prodigiously. Every shepherd knows this, but the actual measurements of this contribution of the vapour-laden air have not often been taken. Yet the subject is an interesting one, and of real importance to all dwellers on high hills, especially those which, like the South Downs, are near the sea, and attract great masses of fog and vapour-laden cloud, but contain few springs on the high rolls of the hills.

The following are some notes of the rise in a dew pond caused by winter fogs on the Berkshire Downs. They were recorded by the Rev. J.G. Cornish at Lockinge, in Berkshire, and taken at his suggestion by a shepherd[1] in a simple and ingenious way. Whenever he thought that a heavy dew or fog was to be expected (and the shepherds are rarely wrong as weather prophets) he notched a stick, and drove it into the pond overnight, so that the notch was level with the surface. Next morning he pulled it up, marked how high the water had risen above the notch, and nicked it again for measurement. On January 18th, after a night of fog, the water rose 1-1/2 in.; on the next day, after another fog, 2 in.; and on January 24th, 1 in. Five nights of winter fog gave a total rise of 8 ins.—a vast weight of water even in a pond of moderate area. Five days of heavy spring dew in April and May, with no fog, gave a total rise in the same pond of 3-1/2 ins., the dews, though one was very heavy, giving less water than the fogs, one of which even in May caused the water to rise 1-1/2 ins.[2] The shepherds say that it is always well to have one or two trees hanging over the pond, for that these distil the water from the fog. This is certainly the case. The drops may be heard raining on to the surface in heavy mists. During the first October mists of 1891 the pavement under certain trees was as wet as if it had been raining, while elsewhere the dust lay like powder. The water was still dripping from these trees at 7 a.m. Under the plane-trees the fallen leaves were as wet from distilled moisture as if they had been dipped in water; yet the ground beyond the spread of the tree was dry. The writer tried a simple experiment in this distilling power of trees. At sundown, two vessels were placed, one under a small cherry-tree in full leaf, the other on some stone flags. Heavy dew was falling and condensing on all vegetation, and on some other objects, with the curious capriciousness which the dewfall seems to show. The leaves of some trees were already wet. In the morning the vessel under the tree, and that in the open, both held a considerable quantity of water, that on the stone caught from dew and condensation, that under the tree mainly from what had dripped from the leaves, which clearly intercepted the direct fall of dew. But the vessel under the tree held just twice as much water as that in the open, the surplus being almost entirely derived from drops precipitated from the leaves. Mr. Sanderson, the manager of the elephant-catching establishment of the Indian Government, noted that in heavy dews in the jungle the water condensed by the leaves could be heard falling like a heavy shower of rain.

Gilbert White, who noticed everything, and lived near a chalk hill, makes some shrewd conjectures, both about the dew ponds and the part which trees play in distilling water from fog, though he does not form the practical conclusion, which we think is a safe one, that the most fog-distilling trees should be discovered and planted to help to supply the water in these air-tapping reservoirs. "To a thinking mind," he writes, "few phenomena are more strange than the state of little ponds on the summits of the chalk hills, many of which are never dry in the most trying droughts of summer. On *chalk* hills, I say, because in many rocky and gravelly soils springs usually break out pretty high on the sides of elevated grounds and mountains; but no persons acquainted with chalky districts will allow that they ever saw springs in such a soil but in valleys and bottoms, since the waters of so pervious a stratum as chalk all lie on one dead level, as well-diggers have assured me again and again. Now we have many such little round ponds in this district, and one in particular on our sheep-down, three hundred feet above my house, and containing perhaps not more than two or three hundred hogsheads of water; yet it is never known to fail, though it affords drink for three or four hundred sheep, and for at least twenty head of large cattle beside. This pond, it is true, is overhung with two moderate beeches, that doubtless at times afford it much supply. But then we have others as small, which, without the aid of trees, and in spite of evaporation from sun and wind and perpetual consumption by cattle, yet constantly contain a moderate share of water, without overflowing in the winter, as they would do if supplied by springs. By my Journal of May, 1775, it appears that 'the small and even the considerable ponds in the vales are now dried up, but the small ponds on the very tops of the hills are but little affected.' Can this difference be accounted for by evaporation alone, which is certainly more prevalent in the bottoms? Or, rather, have not these elevated pools *some unnoticed recruits, which in the night time counterbalance the waste of the day?*" These unnoticed recruits, though it is now certain that they come in the form of those swimming vapours from which little moisture seems to fall, are enlisted by means still not certainly known. The common explanation was that the cool surface of the water condensed the dew, just as the surface of a glass of iced water condenses moisture. The ponds are always made artificially in the first instance, and puddled with clay and chalk.

In the notes to a recent edition of "White's Selborne," edited by Professor L.C. Miall, F.R.S., and Mr. W. Warde Fowler, a considerable amount of information on dew ponds is appended to the passage quoted above, but the source of supply still remains obscure.

The best dew ponds seem to be on the Sussex Downs, where far more fog and cooling cloud accumulates than on the more inland chalk ranges, because of the nearness of the sea. Near Inkpen Beacon, in Hampshire, there is a dew pond at a height of nine hundred feet, which is never dry, though it waters a large flock of sheep.[3] Dew ponds are often found where there are no other sources of supply, such as the wash coming from a road. Probably if the site for one had to be selected, it should be where the mists gather most thickly and the heaviest dews are shed, local knowledge only possessed by a few shepherds. I have driven up *through* rain on to the top of the downs, and found there that no rain was falling, but mists lying in the hollows like smoke. Mr. Clement Reid, F.R.S., has added to the "Selborne" notes his own experiences of the best sites for dew ponds. They should, he thinks, be sheltered on the south-west by an overhanging tree. In those he is acquainted with the tree is often only a stunted, ivy-covered thorn or oak, or a bush of holly, or else the southern bank is high enough to give shadow. "When one of these ponds is examined in the middle of a hot summer's day," he adds, "it would appear that the few inches of water in it could only last a week. But in early morning, or towards evening, or whenever a sea-mist drifts in, there is a continuous drip from the smooth leaves of the overhanging tree. There appears also to be a considerable amount of condensation on the surface of the water itself, though the roads may be quite dry and dusty. In fact, whenever there is dew on the grass the pond is receiving moisture."

Though this is evidently the case, no one has explained how it comes about that the pond surface receives so very much more moisture than the grass. The heaviest dew or fog would not deposit an inch, or even two inches, of water over an area of grass equal to that of the pond. None of the current theories of dew deposits quite explain this very interesting question. Two lines of inquiry seem to be suggested, which might be pursued side by side. These are the quantities distilled or condensed on the ponds, and the means by which it is done; and secondly, the kind of tree which, in Gilbert White's phrase, forms the best "alembic" for distilling water from fog at all times of the year. It seems certain that the tree is an important piece of machinery in aid of such ponds, though many remain well supplied without one.

[1] Thomas Elliot, who for some twenty years was shepherd and general manager for one of my father's tenants at Childrey.

[2] Full details of the cost and method of making dew ponds, as well as other information about them, are contained in the prize essay of the late Rev. J. Clutterbuck, Rector of Long Wittenham, in the Journal of the Royal Agricultural Society. Vol. I., §S. Part 2.

[3] In the Isle of Wight, on Brightstone Downs, about 400 feet above the sea, is a dew pond with a *concrete* bottom, which has never run dry for thirty years.

POISONOUS PLANTS

A friend informs me that he has found a quantity of woad growing on the Chilterns above the Thame, enough to stain blue a whole tribe of ancient Britons, and also that on a wall by the roadside between Reading and Pangbourne he discovered several plants of the deadly nightshade, or "dwale." This word is said to be derived from Old French *deuil*, mourning; but its present form looks very English. The only cases of plant poisoning now common among grown-up people are those caused by mistaking fungi for mushrooms, or by making rash experiments in cooking the former, of which Gerard quaintly says: "Beware of licking honey among the thorns, lest the sweetness of the one do not countervail the sharpness and pricking of the other." But with such a list of toxic plants as our flora can show there is always danger from certain species whose properties are quite unknown to ordinary mortals. Are they equally unknown to the herbalists and that mysterious trade-union of country-women and collectors of herbs by the roadside who deal with them? Probably the trade in poisons not used for serious purposes, but for what used in some parts of England to be called "giving a dose," a punishment for unfaithful, unkind, or drunken husbands, still exists as it did some forty years ago. The collectors of medicinal plants cut from the roadside and rubbish heaps, plants whose "operations" for good are quite well known, and have been handed down by tradition for centuries, cannot be absolutely ignorant of the other side of the picture, the toxic properties which other plants, or sometimes even the same plants, contain. Foxglove, for instance, from which *digitalis* used as a medicine is extracted, is a good example of these kill-or-cure plants. Every portion of the plant is poisonous, leaves, flowers, stalks, and berries. It affects the heart, and though useful in cases in which the pulsations are abnormal, its symptoms when taken by persons in ordinary health are those of heart failure. Thus foxglove is not only a dangerous but a "subtle" poison.

Among other plants which may cause serious mischief, but are seldom suspected, are such harmless-looking flowers as the meadowsweet, herb-paris, the common fool's-parsley, found growing in quantities in the gardens of unlet houses and neglected ground which has been in cultivation, mezereon, columbine, and laburnum. Meadowsweet has the following set against its name: "A few years since two young men went from London to one of the Southern counties on a holiday excursion, on the last day of which they gathered two very large sheafs of meadowsweet to bring home with them. These they placed in their bedroom at the village inn where they had to put up. In the course of the night they were taken violently ill, and the doctor who was called in stated that they were suffering from the poisonous prussic-acid fumes of the meadowsweet flowers, which he said almost overpowered him when he came into the room. The flowers were at once removed, and the young men, treated with suitable restoratives, were by next morning sufficiently recovered to undertake the journey home." [1] Without knowing what the young men had had for supper, it seems perhaps rather hasty to blame the meadowsweet. But the other flowers mentioned above have a bad record. To take them in order. Herb-paris, which grows in woods and shady places, with four even-sized leaves in a star at the top of the stem, all growing out opposite each other, bears a large, green solitary flower, and a bluish-black berry later. All parts of the plant are poisonous, the berries especially. Fool's-parsley, an unpleasantly smelling, very common plant, which leaves its odour on the hand if the seeds are squeezed or drawn through it, is said to cause numbers of deaths by being mistaken for common parsley and cooked. In the case of poisoning by this plant, it is recommended that milk should be given, the body sponged with vinegar, and mustard poultices put on the sufferer's legs. It is reckoned that one plant produced six thousand and eighty seeds—an unpleasant degree of fecundity for a poisonous weed. Columbine, which is a wild plant with blue or white flowers, as well as a domesticated one, has a toxic principle like that of the monkshood, more especially in the seeds; and the pretty red berries of the mezereon are responsible for the deaths or illness of children nearly every autumn. They are like cherries, and easily picked from the low bushes on which they grow. A dozen are said to be enough to cause death, though this must probably depend on the state of the eater's health. The laburnum, with its golden rain, is potentially a kind of upas tree. The writer has only known of two deaths of children caused by eating the beans in the green pods, but it is said to be a frequent cause of death every year on the Continent, where, possibly, children are less naturally careful about poisonous plants than those in England, to whom risks of this kind are usually and properly made part of the "black list" of the nursery-book of "Don'ts." The seeds will even poison poultry, if they pick them up after they have dropped from the pod. Laburnum is of comparatively recent introduction into Britain, or it would probably earlier have been accorded a place among the severely poisonous plants, dreaded by all.

Of these the deadly nightshade and hemlock are the best known in story, while the yew is most dangerous because far more common. In one case the Rector of a Berkshire village was made very ill by eating honey which had been partly gathered from yew flowers. Green hellebore and monkshood are also classed in the list of the ranker poisons. Deadly nightshade is rather a rare plant, yet it may be seen often enough on the sides of woods where there are old walls. It is poisonous throughout. The flowers are large, single, purple bells, and the berries black and shiny like a black cherry. It is said of this dangerous plant that the roots are computed to be five times more poisonous than the berries, that human beings have been found more susceptible to it than animals, and carnivorous animals more so than others. Children suffer more in proportion to the quantity of poison taken than do adults. But cases of nightshade poisoning are very rare, though two were reported some three years ago. Possibly the berries often fail to ripen, and so are less attractive in appearance. The poisonous hemlocks are two, one of which, the common hemlock, is said to have been the plant from which the Athenians prepared their poison for executing citizens condemned to death; and the other, the water-hemlock, or cowbane, is particularly deadly when eaten by cattle, to which it is fatal in a very few hours. Another plant, used for preparing poison in India, which produces a drug used by some tribes of Thugs for procuring the death of their victims, datura or stramonium, has now found a place amongst our wild flowers. It has an English name, thorn-apple, and is said to have been naturalised by the gipsies, who used the seeds as a medicine and narcotic, and carried them about with them in their wanderings. Like henbane, it is often seen on rubbish-heaps and in old brickfields. The leaf is very handsome, and the flower white and trumpet-shaped. Both this plant and the henbane retain their poisonous properties even when dried in hay, and stalled cows have been known to be poisoned by fodder containing a mixture of the latter plant.

Cattle have a delicate sense of smell which warns them of the danger of most poisonous English herbs, though apparently this warning odour is absent from the plants which kill so many horses when the grass grows on the South African veld, and also from our English yew. Yew was anciently employed as a poison in Europe, much as is the curari to-day in Central America. Dr. W.T. Fernie, the author of "Herbal Simples Approved for Modern Use," says that its juice is a rapidly fatal poison, that it was used for poisoning arrows, and that the symptoms correspond in a very remarkable way with those which follow the bites of venomous snakes. It is believed that in India there is a poison which produces the same effect. An Indian Rajah once desired that a notice should be put in a well-known paper that he did not intend to raise his rents on his accession to the estates. The proprietor of the paper asked him his reasons for wishing for such an advertisement. The Rajah said that his grandfather had raised the rents, and had died of snake-bite; that his father had done the same, and had also died of snake-bite; and that he concluded that there was some connection of cause and effect. The notice was inserted, and this Rajah did not die of snake-bite, or rather of the poison which simulates it.

[1] "Farm and Home" Year Book for 1902.

ANCIENT THAMES MILLS

Almost the greatest loss to country scenery is the decay of the ancient windmills and water-mills. The first has robbed the hilltops of a most picturesque feature, while in the valleys and little glens the roaring, creaking, dripping wheel sounds no longer, except in favoured spots where it still pays to grind the corn in the old way. The old town and city mills often survived longer than the country ones, and those on the Thames longer than those on smaller rivers. The corn and barley which was taken to market in the town was easily transferred to the town mill, and thence by water to the place of consumption. Every Wykehamist remembers the ancient and picturesque mills of Winchester, with the mill-stream bridged by the main street. At Oxford some of the most ancient mills remain to this day, while others have only recently been destroyed, or have undergone a curious conversion into dwelling-houses, beneath which the mill-stream still rushes. One of these houses stands near Folly Bridge; another old mill has just undergone the same process, that close to Holywell Church. Some of these mills are the most ancient surviving institutions in Oxford, far older than the colleges—older even than any of the churches except perhaps one. Some of these—the Castle Mill, for instance—have ground corn for centuries since the abbeys, for whose use they were founded, utterly disappeared. Others were standing long before abbeys or colleges were founded, and were part of their endowments. They are the oldest link between town life and country life left in Oxford, or indeed in England. For a thousand years the corn grown on the hills beyond the Thames meadows has been drawn to their doors. Saxon churls dragged wheat there on sledges, Danes rowed up the river to Oseney and stole the flour when they sacked the abbey, Norman bishops stole the mills themselves. That iniquitous Roger of Salisbury was "in" this, as we might guess. Roger, who knew that attention to detail is the soul of business, commandeered this particular mill with others in these parts, and, when forced to let it go, with a fine sense of humour made it over to the Godstone nunnery as a pious donor.

The Knights Templars had another mill at Cowley, and the king himself one on the Cherwell, which was given to the Hospital of St.

John, who "swapped" it with Merton. Later on these mills helped King Charles's army vastly, for all the flour needed for the Oxford garrison was ground inside or close to the walls.

At present the Thames is mainly visited as a source of rest and refreshment to tens of thousands of men "in cities pent," and of pleasure rather than profit. In a secondary degree it is useful as a commercial highway, the barge traffic being really useful to the people on its banks, where coal, stone for road-mending, wood, flour, and other heavy and necessary goods are delivered on the staithes almost at their doors. But when the old mills were first founded, and for eight centuries onwards, it was as a source of power, a substitute for steam, that the river was valued. The times will probably alter, and the Thames currents turn mill wheels again to generate electric light for the towns and villages on its banks. The chance of this coming about is enough to make any one who owns a mill right on the water keep it, even though not useful at present. First the old roads with auto-cars, then the old mills with hydraulic lighting and low-power dynamos will come to the front again. Whereof take the old story of the Oxford river as full and sufficient witness, and Antony Wood for storyteller. "Oxford," he says, "owed its prosperity to its rivers," of which there were apparently as many branches and streams then as now.

The rivers were "beneficial to the inhabitants, as anon shall be showed," though the Cherwell was "more like a tide" than a common river sometimes, and once nearly overflowed all the physic garden. That garden stands there still. So does the Cherwell still behave "more like a tide than a river," and the scene at the torpid races a few years ago is evidence that the rivers have not diminished in volume. What, then, was the "great commodity" given by them to the city? First and least, a water which was good for dyeing cloth and for tanning leather; secondly, and by far the greatest benefit, it turned the wheels of at least a dozen important mills. As mills were always a monopoly, as much opposition was raised to the making of a new one as would now be evoked by the proposal to construct a new railway.

It was meddling with vested interests of a powerful kind, but there were so many rivers at Oxford that each turned one or two mills without injuring any one's water rights.

Of all these mills, the greatest advantage to the city came from the Castle Mill. Notwithstanding its name, this was *not* the property of the Castle of Oxford, though it stood within arrow-shot of its towers, and was thus protected from pillage in time of war. It stands under the remaining tower, the water tower, of the castle still, and on exactly the same site, and on the branch of the Thames which from the most ancient days has been the waterway by which barges and merchandise came from the country to the city, bringing goods from Abingdon or corn and fuel from the upper river. And it is still called by its old name of the Weir Stream. "There is one river called Weyre, where hath bin an Hythe, at which place boatmen unload their vessels, which also maketh that antient mill under the castle seldom or never to faile from going, to the great convenience of the inhabitants." So says Antony Wood, adding that it stood before the Norman conquest. After that it was forfeited to the Norman kings, and then held in half shares by the burgesses of the town and the abbots of Oseney, that once wealthy and now vanished abbey, which stood close by where the railway station now is. They shared the fishery also, and apparently this partnership prevented friction between the town and the monks, as each could undersell the other, and prices for flour and fish were kept down at a reasonable figure.

Henry VIII. gave the abbey's share to the new bishopric of Oxford, but the funds of the bishopric were embezzled by some means, and the town ultimately bought the mill for £566.

St. George's Tower, the only remaining fragment of the castle, is built of stones and mortar, so compact that though the walls have stood since Robert d'Oily reared it, late in the reign of the Conqueror, the stones and mortar had to be cut out as if from a mass of rock when a water-pipe was recently taken through the walls. It is now the water tower which holds the supply for Oxford prison.

Old Holywell Mill was on a branch of the Cherwell, and stood just behind Magdalen Walks, whence a charming view was had of its wheel and lasher. It belonged to the Abbey of Oseney, who gave it to Merton College in exchange for value. Now it is a handsome dwelling-house, below which the mill stream rushes.

[Illustration: BOTLEY MILL. *From a photograph by Taunt & Co.*]

[Illustration: EEL BUCKS. *From a photograph by Taunt & Co.*]

Merton College seems to have had a fancy for owning mills, for it also acquired by exchange the King's Mill. Only the house and lasher are left to show where this old mill stood. It had a narrow but very strong mill stream, which in winter used to come down in a sheet of solid water like green jade, a beautiful object among the walks and willows of Mesopotamia. It was an outpost of the King's forces when Oxford was held for the Royalists.

Botley Mill, though on the westernmost of the many streams into which the Thames divides at Oxford, was outside the walls. It dates from before the Conquest. This belonged to the Abbey of Abingdon, in the chronicles of which are some records of an injury done to the "aqueduct, which is vulgarly called the lake." This name is still the local term for all side streams and artificial cuts from the Upper Thames. The men of a now vanished village of Seckworth broke the banks of the "lake" when Odo, Bishop of Bayeux, was being besieged in Rochester Castle. The lord of the manor was subsequently sued for this by the abbot of Abingdon, and had to pay ten shillings damages. Doubtless the men of Seckworth had to contribute to pay for their indulgence in this mischief, but it looks as if the abbot's miller had been cheating them.

THE BIRDS THAT STAY

In the Vision of the Lots and Lives, when the souls chose their careers on a fresh register before taking another chance in the world above, Ulysses chose that of a stay-at-home proprietor, with a resolve, born of experience, never again to roam. If Plato had made a Myth of the Birds, he might have alleged some such reason to explain how it is that while most of them are incessant wanderers, ever flitting uncertain between momentary points of rest, so few remain fixed and constant, as if they had sworn at some distant date never more to make trial of the wine-dark sea. In the still, November woods, when the vapours curl like smoke among the dripping boughs, leaving a diamond on each sprouting bud where next year's leaf is hid; by the moorland river, on bright December mornings, when the grayling are lying on the shallows below the ripple where the rock breaks the surface; by the frozen shore where the land-springs lie fast, drawn into icicles or smeared in slippery slabs on the cliff faces, and hoar frost powders the black sea-wrack; on the lawns of gardens, where the winter roses linger and open dew-drenched and rain-washed in the watery sunbeams—there we see, hear, and welcome the birds that stay. Then and there we note their fewness, their lameness, and feel that they are really fellow-countrymen, native to the soil. The list of these home-loving birds is short; and those commonly seen are only a few of the total. In a winter stroll by the upper Thames, the absence of the birds which flocked along the banks in summer and spring, when the May was in blossom and the willow covered with cotton fleck, is among the first seasonal changes noticed. The chiff-chaffs, turtledoves, sedge-warblers, whitethroats, coots, sandpipers, and all the little river birds are gone. So are the greater number of the blackbirds, thrushes and missel-thrushes. All the fisherman sees, his daily companions by the deserted river, are the wren creeping in the flood-drift, the tits working over the alder bushes to see if any seeds are left in the cones, and the kingfishers. The grayling fisherman on the Northern streams has the water ousels for his constant and charming companions, true to the mountain river as in the days of Merlin and Vivien, busy as big black-and-white bees as they flit up-stream and down-stream, flying boldly into the waterfalls, dropping silently from mossy stones into the clear brown eddies, singing when the sunbeams shine and warm the crag-tops, and even floating and singing on the water, like aquatic robins. The ousels must have been the sacred birds of Tana, the Water Goddess, the ever attached votaries of her dripping and rustic shrines.

By the winter shore, untrodden by any but the fisher going down at the ebb to seek king-crab for bait, or by his children, gathering driftwood on the stones, one little bird stays ever faithful to the same short range of shore. This is the rock-pipit—the "sea-lark" of Browning's verse. But that is a summer song. It is not only when the cliff—

"Sets his bones,
To bask i' the sun,"

but in the short winter days, that the sea-lark keeps constant to the fringe of ocean. It is the most narrowly local and stay-at-home of all birds, never leaving the very fringe and margin, not of sea, but of land, haunting only the last edge and precipice of the coast, nesting on those upright walls of granite or chalk, and creeping, flying, and twittering among the crumbling stones, the water-worn boulders, and the tufts of sea-pink and samphire. When the winter storms slam the roaring billows against the cliff faces and the spray flies up a hundred feet from the exploding mass, the little sea-larks only mount to higher levels of the cliff, never coming inland or forsaking its salt-spattered resting-place. Compared with these home-loving birds, all the gulls are wanderers, even though they do not desert our shores and come fifty miles up the Thames. Of the rock-fowl, the puffins fly straight away to the Mediterranean, and the guillemots and razorbills go out to sea and leave their nesting crags. Only the cormorants stay at home, flying in to roost on the same lofty crag every autumn and winter night, from the fishing grounds which the sea-crows have frequented for longer years even than the "many-wintered crow" of inland rookeries has his fat and smiling fields.

The discovery that rooks, with their reputation for staunch attachment to locality, are regular and irrepressible migrants, crossing from Denmark and Holland to England, and from England to Ireland, has been followed by other curious revelations about the mobility of what were believed to be stationary birds. Our own beloved garden robin, whom we feed till he becomes a sturdy beggar, though he pays us with a song, stays with us, as we know, because he applies regularly for his rations. But he sends all his children away to seek their fortunes elsewhere, and on our coasts flights of migrant robins, whom either their parents, or the bad weather, have sent from Norway over the foam, arrive all through the autumn. Even the jenny-wrens migrate to some extent.

Because we see birds of certain kinds near our farms, gardens, and hedges it does not follow that these are those which were there in summer and spring. Such common finches as the greenfinches and chaffinches migrate in immense flocks, and over vast distances, considering their short wings and small size. In the gardens and shrubberies round the houses the parent robins stay. So do some of the blackbirds, the thrushes (except in very hard weather), the hedge-sparrow, the nuthatch (more in evidence in winter than at any other time, and a firm believer in eleemosynary nuts), all the tits, except the long-tailed tit, a little gipsy bird wandering in family hordes, and the crested and marsh tits (dwellers in the pine forest and sedge-beds), and the wood pigeon. Occasionally that shy bird, the hawfinch, is seen on a wet, quiet day picking up white-beam kernels and seeds. Except this, every one of the garden birds comes to be fed, and is well known and appreciated. It is in the woods and the hedges of the rain-soaked meadows that the general absence of bird life in winter is most marked, and the presence of the few which stay most appreciated. Those who, on sport intent, go round the hedges in November and December, or wait in rides while the woods are driven, or lie up quietly in the big covers for a shot at wood pigeons in the evening, are almost startled by the tameness and indifference of the birds, eagerly feeding so as to make the most of the short, dark days. When the hedges are beaten for rabbits the bullfinches appear in families, their beautiful grey backs and exquisite rosy breasts looking their very best against the dark-brown, purply twigs. Another home-staying bird of the hedgerows, or rather of the hedgerow timber, is the tree-creeper. It has no local habitation, being a bird which migrates in a drifting way from tree to tree, and so bound by no ties to mother-earth. But it is in the woods that the stay-at-home birds are most in evidence in winter. There they find abundant food, and there they make their home. The woodpeckers, the magpie, and the jay, the brown owl, the sparrow-hawk, the kestrel, the pheasant, the long-tailed tit, and all the rest of the tribe; and in the clearings where the teazle grows, the goldfinches feed. The barn owl and brown owl both stay with us. So does the long-eared owl. But the short-eared owl is a regular migrant, coming over in flights like woodcock. No one has satisfactorily answered the question why there are sedentary species and migratory species so closely allied in habits and food that the quest for a living must be ruled as outside the motive for migration.

If the long-eared owl can remain and find a living all the year round in the copses on the downs, why should not the short-eared owl make a practice of what is its occasional custom, and nest in the fens and marshes? If the kingfisher can find a living and abundant fish in our rivers and brooks, why does the dabchick migrate? The migration is only a partial one, for many remain on the Thames all the year round, especially near the eyots by Tilehurst; but it vanishes from most of the Northern pools and returns almost on the same date. Perhaps a conclusion might be hazarded from the behaviour of wild migratory birds which have become semi-domesticated. In Canada, the largest and best known of the wild geese is the black-necked Canadian goose. It is a regular migrant. The Indians believe it brings little birds on its back when it comes. At Holkham, where a large flock of these is acclimatised, but lives under perfectly wild conditions, the Canadian geese never attempt to migrate, though they often fly out on to the sands at ebb-tide. They show less disposition to leave the estate than the herons in the park. Yet during the winter they feed every day with flocks of wild geese in the marshes. These geese fly every spring away to the Lapland mountains or the tundras, and could show the Canada geese the way northwards if they wished to follow. The conclusion is that the Canada geese have no desire for change; and the reason that other birds do not migrate is probably the same.

ANCIENT HEDGES

In the upper Thames valley, both in May and autumn, one of the prettiest sights is the great hedges which divide the meadows. In spring, those above Oxford look as though covered with snow, and in early October they are loaded with hips and haws, just turned red, with blackberries, elderberries (though the starlings have eaten most of these), with crab apples, with hazel nuts, scarlet wild guelder-rose berries, dog-wood berries, and sloes. Except the fields themselves, our hedges are almost the oldest feature with which Englishmen adorned rural England. They have gone on making them until the last parish "enclosures," some of which were made as late as thirty years ago, and when made they have always been regarded as property of a valuable kind. When Christ's Hospital was founded in Ipswich in Tudor days, partly as a reformatory for bad characters, "hedge-breakers" were more particularly specified as eligible for temporary domicile and discipline. "Hedges even pleached" were always a symbol of prosperity, care, and order. "Her fruit trees all unpruned, her hedges ruined," a token that something was amiss in our country economy.

One untidy habit, which the writer remembers as very common, has been discontinued in this connection. Twenty years ago the linen drying on the hedge, which Shakespeare evidently regarded as a "common object of the country," was constantly seen. It was always laid on well-trimmed hedges, or otherwise it would have been torn. Now it is always hung on lines, possibly because the hedges are not so well trimmed and kept. Bad times in farming have greatly helped the beauty of hedges. They are mostly overgrown, hung with masses of dog-rose, trailed over by clematis, grown up at bottom with flowers, ferns, and fox-gloves, festooned with belladonna, padded with bracken. The Surrey hedges are mostly on banks, a sign that the soil is light, and that a bank is needed because the hedge will not thicken into a barrier. But these, like most others, are set with the charming hedgerow timber that makes half England look like a forest at a distance of a mile or so. It is difficult to reconstruct our landscape as it was before the hedges were made. But any one curious as to the comparative antiquity of the fields can perhaps detect the nucleus or centre where enclosure started. Those having the ditch on the outer side are always the earlier, the ditch being the defence against the cattle that strayed on the unenclosed common or grazings outside.

The finest garden hedges in England are at Hall Barn, in Buckinghamshire. They must be thirty feet high, are immensely thick, and are clipped so as to present the smooth, velvety appearance peculiar to the finest yew and box hedges. The colour and texture of these walls of ancient vegetation, contrasting with the vivid green lawns at their feet, are astonishingly beautiful. One of the peculiar charms of such hedges is that where yew of a different kind or age, or a bush of box, forms part of the mass, it shows like an inlay of a different material, and the same effect is given merely by the trick that some yews have of growing their leaves or shoots at a different angle from that favoured by others. These surfaces give the variety of tint which is shown in such fabrics as "shot" or "watered" silk. Here there is a splash of blue from the box, or of invisible dull green, or of golden sheen, from different classes of yew. Box hedges of great size are less common than those of yew, and less durable, for the box is easily rent from the stem when old. But these two, the yew and the box, are the "precious" hedges, the silver and gold, of the garden-maker. Next, representing the copper and brass, are the hedges of beech and holly. Both are commonly planted and carefully tended as borders and shelters to the less important parts of gardens; as screens also to block out the humdrum but necessary portions of the curtilage, such as the forcing-pits for early plants, minor offices, timber yards, and the like; and to shelter vegetable gardens (for which the Dutch use screens of dried reeds). Holly makes the best and most impenetrable of all hedges when clipped, but it is not beautiful for that reason. Clipped holly grows no berries; it accumulates dust and dirt, and has a dull, lifeless look. Beech, on the other hand, should be in greater esteem than it is. If clipped when the sap is rising it puts on leaves which last all the winter. From top to bottom the wall of russet shines warm and bright. Its leaves are harmless in decay, for they contain an antiseptic oil, and no leaves of spring are more tenderly green or in more ceaseless motion at the lightest breeze. Privet makes the last and least esteemed of these "one-tree" hedges. Yet it is the most tractable of all hedge material, and was almost invariably used to form the intricate "mazes," once a favourite toy of the layers-out of stately gardens.

Keeping these hedges in good repair and properly clipped and trimmed is one of the minor difficulties of the country. In large gardens there are always one or two professional gardeners who understand the topiary art. But it often happens that a quite modest garden possesses a splendid hedge of yew or box, the pride of the place, which needs attention once or twice every year. These hedges have frequently been clipped by the same man, some old resident in the village, for thirty or forty years. Clipping that hedge is part of his regular extra earnings to which he looks forward, and a source of credit and renown to him in his circle. He knows every weak place, what parts need humouring, what stems are crowding others between the furry screen of leaves, and where the wind got in and did mischief in the last January gale. When in the course of Nature the old hedge-trimmer dies, there is no one to take his place. The men do not learn these outside accomplishments as they once did, and the art is likely to be lost, just as ornamental thatching and the making of the more decorative kinds of oak paling are in danger of disappearing.

Mending, or still worse remaking, field-hedges is a difficult, expensive, and withal a very highly skilled form of labour. The workers have for generations been very humble men, who have scarcely been honoured for their excellent handiwork as they deserved. They appear in art only in John Leech's pictures of hunting in Leicestershire, in his endless jokes on "mending the gaps" towards the close of the hunting season. In February and March the scenes shown in Leech's pictures are reproduced on most of the Thames valley farms in Berkshire and Oxfordshire. The men wear in front an apron of sacking, torn and plucked by thorns. The hands are gloved in leather mits with no fingers; in them the hedger holds his light, sharp billhook, shaped much like the knife of the forest tribes of Southern India. When a whole fence has to be relaid the art of "hedge carpentry" is exhibited in its perfection. Few people not brought up to the business, which is only one minor branch of the many-sided handiness of a good field labourer, the kind of man whom every one now wants and whom few can find, would have the courage to attempt it. A ditch full of brambles, often with water at the bottom, has to be cleared. Then the man descends into the ditch, and strips the bank of brambles and briars. That is only the preliminary. When he has piled all the brambles in heaps at regular intervals along the brow of the ditch, he walks thoughtfully from end to end of the fence, and considers the main problem, or lets the idea sink into his mind, for he never talks, and probably never frames for himself any form of words or conscious plan. In front, with the bases of the stems bare where the bank is trimmed and slashed, stands the overgrown hedge which he is to cut, bend over, relay, and transform, to make another ten or twelve years of growth till it reaches the unmanageable size of that which stands before him. Most of it is great bushes of blackthorn, hard as oak, with thorns like two-inch nails, and sharper. These bushes, grow up in thick rods and stocks, spiny and intractable, from the bank to a height of perhaps twelve feet. The rest of the fence-stuff is whitethorn, nearly as ill to deal with as the blackthorn, and perhaps a few clumps of ash and wild rose. Slashing, hewing, tearing down, and bending in, he works steadily down the hedge day by day. All the time he is using his judgment at every stroke. Some he hews out at the base and flings behind him on the field. Much he cuts off at what will be the level of the hedge. But all the most vigorous stems of blackthorn and whitethorn he half cuts through and then bends over, twisting the heads to the next stocks or uprights, or, where there are no stocks, driving in stout stakes cut from the discarded blackthorns. When finished the newly mended hedge consists of uprights, mostly rooted in their native bank, and fascine-like bundles—the heads of these uprights, which are bent and bound horizontally to the other uprights or stakes. This is the universal "stake and bond" hedge of the shires, impenetrable to cattle, unbreakable, and imperishable, because the half-cut bonds, the stakes, and the small stuff all shoot again, and in a few years make the famous "bullfinch" with stake and bond below, and a tall mass of interlacing thorns and small stuff above.

During the last era of prosperous farming there was a mania for destroying hedges and cutting down the timber. If ever prosperity returns it will smile on a better-informed class of occupier and owner. It is now seen that the hedges were of the greatest value to shelter cattle, sheep, and horses, and benefited to some extent even the sown crops, especially at the blossoming time. As cattle are now the farmer's main reliance, it will be long before he grubs up or destroys the welcome shelter given by the hedges from sun, rain, and storm.

THE ENGLISH MOCKING BIRD

One winter an unusual number of peewits visited the flats near Wittenham and Burcote, and remained there for several months. One or two starlings which haunted the house in which we stayed, and slept in their old holes in the thatch, picked up all the various peewits' calls and notes, and used to amuse themselves by repeating these in the apple-trees on sunny mornings. The note was so exact a reproduction that I often looked up to see where the plover was before I made out that it was only the starling's mimicry.

A correspondent of the *Newcastle Journal*, writing from Yeare, near Wooler, in Northumberland, recently described the performances of a wild starling which has settled near his house. It is such an excellent mimic of other birds' notes that no one can help noticing its performances. A record has been kept of the variety entertainments provided by the bird. Besides its own calls, whistles, and song, it reproduces the song of the blackbird and thrush absolutely correctly, and mimics with equal nicety the calls of the curlew, the corncrake, and the jackdaw.

It is appropriate that this eulogy of the starling should appear in a Newcastle paper, for Bewick when residing there always regretted the absence of these birds from the town, and hoped that they might in time become numerous, as in the South and West. Starlings are such intelligent, interesting, and really remarkable birds that if they were rare they would be among the most prized of pets. Their open-air vocal performances are quite as remarkable as their latest admirer says. They are the British mocking-birds, able, when and if they choose, to reproduce almost any form of song. They do this partly, no doubt, because their throats are adaptable, but more from temperament and a kind of objective mind not very common in birds. Like parrots, starlings are given to spending a good deal of every fine morning in contemplating other people, including other birds, and then in thinking them over, or talking them over to themselves. Any one who is sitting or working quietly near a room where a parrot is in its cage alone can fairly follow the train of thought in the parrot's mind. It is evidently recalling episodes or things which form part of its daily mental experiences. It begins by barking like the dog, then remembers the dog's mistress, and tells it to be quiet, as she does. Then it hears the housemaid, and imitates a window-sash being let down, or some phrase it has picked up in the servants' quarters. If it has been lately struck with some new animal noise or unusual sound, it will be heard practising that. Starlings do exactly the same thing. When the sun begins to be hot on any fine day, summer or winter, the cock bird goes up usually alone, to a sunny branch, gable, or chimney, and there indulges in a pleasant reverie, talking aloud all the time. Its own modes of utterance are three. One is a melodious whistle, rather low and soft; another is a curious chattering, into which it introduces as many "clicks" as a Zulu talking his native language; and the third is a short snatch of song, either its own, or one which has become a national anthem or morning hymn common to all starlings, though it may originally have been a "selection" from other birds' notes. Then, or amongst the rest of the ordinary notes, the starling inserts or practises its accomplishments. Not all starlings do this, and only a few attain great eminence in that line. Obviously it is only personal feeling that induces them to do it, and they get no encouragement from other starlings, though when kept in cages, as they very seldom are now, and rewarded and taught, they might develop the most striking talents. It should be added that, like all good bird-mimics, they are ventriloquists. They can reproduce perfectly the sound of another bird's note, not as that bird utters it, but as it is heard, faint and low, softened by distance. They can also sing over bars of bird-songs in a low tone perfectly correctly, and repeat them in a high one.

To give a rather striking example. Last spring the writer was in the Valley of the Eden, opposite Eden-hall. The vale is a wide one, and on the north-east side are high fells, Cross Fell among others. On these the curlews breed, and occasionally fly right over the valley at a great height to the hills above Edenhall, uttering their long, musical call. When heard, this call is generally uttered several hundred feet above the valley. A curlew was heard flying above, and repeating its cry, but was not discernible. Again the call was heard, but no curlew seen, though such a large bird must have been visible. In the line of sound was a starling sitting on a chimney-pot. Again the curlew called, the long-drawn notes sounding from exactly the same place in the sky. It was the starling, reproducing with perfect accuracy the call, as it was used to hear it from the highflying curlews crossing the valley. Apparently the tradition that they were good talkers has died out in rural England. It was always one of the firm beliefs of East Anglia that if a starling's tongue were slit with a thin sixpence it would learn to talk at once, but that otherwise it would only mimic other birds. The operation, like most other traditional brutalities, was absolutely unnecessary. Talking starlings were common enough, and must have been for many years previous to the time when they were no longer valued as cage-birds. Has not Sterne in his "Sentimental Journey" immortalised the poor bird whose one and leading sentiment, had he been able to find words for it, was "I can't get out! I can't get out!"?

* * * * *

From early spring until after midsummer the starlings have young broods in more varied places and positions than probably any other birds in England. They like the homes of men, and build with equal pleasure in thatched roofs, under tiles, in the eaves and under the leads of churches (though a recent edict by the Bench of Bishops has forbidden them the towers by causing wire netting to be placed over the louvre boards), and also in places the most remote from mankind. In the most solitary groves on Beaulieu Heath, under the ledges of stark Cornish precipices, and in ruins on islets in mountain lochs in Scotland, they tend their hungry nestlings with the same assiduous care. The good done by the starlings throughout the spring, summer, and autumn is incalculable. The young are fed entirely on insect food, and as the birds always seek this as close to home as possible, they act as police to our gardens and meadows. They do a little mischief when nesting and in the fruit season, partly because they have ideas. It was alleged recently that they picked off the cherry blossoms and carried them off to decorate their nests with. Later they are among the most inveterate robbers of cherry orchards and peckers of figs, which they always attack on the ripest side. But they have never developed a taste for devouring corn, like the rice-birds and starlings of the United States. They have a good deal in common with those bright, clever, and famous mimics, the Indian mynahs, which they much resemble physically. This was the bird which Bontius considered "went one better" than Ovid's famous parrot: —

"Psittacus, Eois quamvis tibi missus ab oris
Jussa loquar; vincit me sturnus garrulus Indis."

The mynahs have also the starling's habit of building in houses, and especially in temples. There is a finish about the mynah's and the starling's mimicry which certainly beats that of the parrots.

In their attendance on sheep and cattle the starlings have another creditable affinity. They are very like the famous rhinoceros-birds of Africa, to which also they are related. The rhinoceros-birds always keep in small flocks, every member of which sits on the back of the animal, whether antelope, buffalo, or rhinoceros, on which it is catching insects. The starlings do not keep so closely to the animal's body, though they frequently alight on the back of a sheep or cow and run all over it. But when seeking insect food among cattle the little groups of starlings generally keep in a pack and attend to a single animal. Mr. J.G. Millais, watching deer in a park with his glasses, saw a starling remove a fly from the corner of a deer's eye. When they have run round it, and over it, and caught all the flies they can there, they rise with a little unanimous exclamation, and fly on to the next beast. Their winter movements are also interesting. By day they associate with other birds, mainly with rooks. Gilbert White thought they did this because the rooks had extra nerves in their beaks, and were able to act as guides to the smaller birds searching for invisible food. Probably it is only due to the sociable instinct. Towards night they nearly always repair in innumerable flocks to some favourite roosting-place, either a reed-bed or a wood of evergreens, where they assemble in thousands. One of these communal sleeping-places is the duck island in St. James's Park. In hard weather they feed on the saltings and round the shore, especially where rotten seaweed abounds, with great quantities of insect life in it. At such times they roost in the crevices of the great sea cliffs. Under Culver Cliff, for instance, they may be seen flying along the shore and coming in to bed in the frost fog with the cormorants and other fishers of the deep.

FLOWERS OF THE GRASS FIELDS

Just before hay-time, the crowning glory of the Thames-side flats is given by the flowers growing in the grass. Their setting, among the uncounted millions of green grass stems, appeals not only by the contrast of colour, but by the sense of coolness and content which these sheltered and softly bedded blossoms suggest. The meadows which they adorn are best-loved of all the fields of England; but they would never be as dear to Englishmen as they are were it not for the flowers which deck them. The blossoms and plants found in the tall grasses differ from those on lawns and grazing pastures. They are taller, more delicate, and of a more graceful growth. The daisy, so dear to pastoral poets, is not a flower of the hayfield. The myriads of springing stems choke the daisy flowers, which love to lie low, on their flat and shallow-rooted stars of leaves. The daisy is a lawn plant that loves low turf, and only in early spring on the pasture-fields does it whiten the unmown grasses. The turf glades of the New Forest, grazed short by cattle for eight hundred years, are very properly called "lawns"; and on these the daisies grow in thousands, showing that they are true lawns, and not grassfields mown yearly by the scythe. What makes a flower of the grasses it is difficult to say. Bulbs flourish among them, and clovers, trefoils, and vetch. White ox-eye daisies love the grass, and many orchids, and in shady places white cow-parsley, and blue wild geraniums, and all the buttercups. Others, like the yellow snapdragon and the scarlet poppy, will have none of it, but love a dry and dusty fallow or a cornfield that has run to waste, shimmering with heat and drought. Up the valley of the Pang, you may see acres of poppies on a fallow as scarlet as a field-marshal's coat, and not one in the meadows by the stream. Even before the sheltering grass stems shoot upward and around them, drawing all the flower-life skywards as trees draw other trees upright towards the light, there are plants which are found only growing in the meadows, springing from the turf carpet, and happy in no other setting. Chief of these are the wild daffodils or Lent-lilies, the ornaments of old orchards and of the green meadows of Devon and the Isle of Wight. Why they, like the snowdrops, and in other parts of Europe the narcissi, should choose the turf in which to flower, instead of the woods, where grass does not grow, is one of the secrets of the flower-world. So, too, the wild hyacinths grow not in the meadows, though the fritillaries, the chequered red or pale "snake flowers," are grass-lovers, and grow only in the alluvial meadows by the streams and brooks of the valleys. Early though the fritillaries are, they are a real "grass flower," flourishing best where there is some early succulent growth around them, for they like the shelter so given. This they enjoy even early in the year, because their favourite home is in meadows over which flood-waters run in winter, and there the grass grows fast. With the cowslip comes the early common orchis, with its red-purple flower, and later the masses of buttercups, and the ox-eye daisies. Both these flowers are increasing in our meadows, the former to the detriment of the grass itself, and to the loss of the butter-makers, for the cows will not eat the buttercups' bitter stems. Like the ox-eye daisy, the buttercup is a typical meadow flower, tall, so that it tops the grasses and catches the sun in its petals, thin-foliaged, for no real grass-growing flower has broad or remarkable leaves, and with a habit of deep, underground growth far below the upper surface of the matted grass roots. You cannot easily pull up a buttercup root, or that of any flower of the meadows. The stems break first, for they draw their sustenance from a deep stratum of earth. Most of the meadow flowers and blossoms in the mowing grass belong to the beautiful, rather than to the useful, order of plants. They are fitted to weave a garland from rather than to distil into simples and potions. As Gerard says of the butterfly orchis, "there is no great use of these in physicke, but they are chiefly regarded for the pleasant and beautiful flowers wherewith Nature hath seemed to play and disport herselfe." Herein they differ from the roadside plants and the blossoms of waste-lands and woods, for these, especially the former, swell the list of the medicinal plants, the garden not of Flora, but of Aesculapius. It is these which have been gathered for centuries by the wise men and wise women of the villages from the Apennines to Exmoor, while, if we may infer from the story of agriculture, the flowers of the grass-fields are in a sense modern and artificial. They owe their numbers to the discovery of the art of haymaking. Before men learnt to cut, dry, and stack hay, which, after fermenting partly in the stacks under pressure, becomes a manufactured food, it may be concluded that there were no such flower-spangled fields, in this country at least, as now form such a striking feature of rural England. Cattle and sheep wandered all over the common pastures, and ate the grass down, or trampled it under foot. Consequently, it never grew long, or formed the protecting bed in which the flowers now lie, and many of the meadow plants could seldom have flowered at all. The hungry cattle would graze down all the soft, juicy young buds and leaves, wandering at will over the valleys, under charge only of the herdsman. When haymaking became general the cattle were confined in spring and early summer, and the fields of "mowing grass" appeared, and nourished year by year the plants peculiar to this form of cultivation. The proof that this is so may be seen in the New Forest. There the private fields, carefully protected during the spring, from the tread or bite of cattle, and mown yearly in the summer, have all the wealth of flowers peculiar to our hay-meadows. Outside, in the forest itself, these flowers hardly exist, except by some pool-side, or on the meadow-like border of a bog. They are only natural in the second sense, because our mowing grass is a natural product of enclosed ground, when cattle are excluded. Some flowers just invade the meadows, venturing out a few yards from the hedges or woods, but never spreading broadcast over the sun-warmed central acres. Such are the blue bird's-eye, which just colours the mowing grass in shady spots and patches near the fence, and occasionally the bee-orchis and the butterfly-orchis. The latter does not grow tall in the meadows as it does in the woods, but affects a humbler growth. Blue wild geraniums also flourish in patches in the meadows, and sometimes cranesbill and campion. But campions do not seed well among the thick grasses and seldom hold their own, as they do where a copse has been cut down, or on a hedgeside. And, though it is not a flower, there is the "quaking grass" beloved of children, though useless as cattle food, and a sign of bad pasturage, but the only grass which cottage people gather to keep, as a memento of the hayfields.

[Illustration: ORCHIS. *From photographs by E. Seeley.*]

Flowering plants form a large part of the actual herbage from which the hay is made. The bottom of a good crop of mowing grass springs from a tangle of clover and leguminous plants, all owning blossoms, and many of them of brilliant hues and exquisite perfume. Chief among these is the red meadow-clover, the pride of the hayfields. Few plants can match its perfume, or the cool freshness of its leaves. With this is mixed the little hop-clover, and the sucklings, and other tiny gold-dust blossoms. Meadow vetchling, and the tall meadow crowfoot, with rich yellow blooms and dainty leaves, are set off by the pinks of the clover and the crimson of stray sainfoin clusters. All these blossoms with the various flowers of the grasses, tend to ripen and come to perfection together, the heats of June bringing the whole multitude on together as in a natural forcing-pit. It is then that the mowing grass is said to be "ripe," when all the blossoms are shedding their pollen, and giving hay-fever to those who enter the fields. It must be cut then, wet or fine, or the quality and aroma of the hay passes away beyond recovery. Perhaps it is an accident that most of our meadow flowers are white or yellow. The two most striking exceptions are from foreign soil, the purple-blue lucerne and the crimson sainfoin. But yellow is not the universally predominant hue of the flowers of grasses, for in Switzerland and the Italian Alps the hayfields are as blue with campanulas as they are here yellow with buttercups. The turf on our chalk downs shows flowers more nearly approaching in tint the flora of the Alps. The hair-bells with their pale blue, and the dark-purple campanulas, give the complement of blue absent in the lower meadows, while the tiny milkwort is as deep an ultramarine as the Alpine gentians themselves. But the turf of the chalk downs, never rising to any height, and without the forcing power of the valley grasses, yields no such wealth of colour or perfume as the meadow flowers lavish on our senses in the early weeks of June.

RIVERSIDE GARDENING

"And a river went out of Eden to water the garden."

A Recent addition to the country house is the "water garden," in which a running brook is the centre and *motif* of the subsidiary ornaments of flowers, ferns, trees, shrubs, and mosses. Nature is in league with art in the brook garden, for nowhere is wild vegetation so luxuriant, and the two forces of warmth and moisture so generally combined, as by the banks of running streams. The brook is its own landscape gardener, and curves and slopes its own banks and terraces, sheltered from rough winds and prone to the sun.

Many houses near the Thames, especially those under the chalk hills which fringe much of the valley, have near them some rill or brook running to the main river. On the sides of the chalk hills, though not on their summits, these streams cut narrow gullies and glens. Wherever, in fact, there is hilly, broken ground, the little rills form these broken ravines and gullies, often only a few yards in width from side to side. Usually these brooklet valleys are choked with brambles or fern, and filled with rank undergrowth. Often the stream is overhung and invisible, or dammed and left in soak, breeding frogs, gnats, and flies. The trees are always tall and beautifully grown, whatever their age, for the moisture and warmth force vertical growth; the smaller bushes—hawthorn, briar, and wild guelder-rose—also assume graceful forms unhidden, for they always bow their heads towards the sun-reflecting stream. Part of the charm of the transformation of these brookside jungles into the brookside garden lies in the gradual and experimental method of their conversion. Every one knows that running water is the most delightful thing to play with provided in this world; and the management of the water is the first amusement in forming the brook garden. When the banks have been cleared of brambles to such a distance up the sides of the hollow as the ground suggests, and all poor or ill-grown trees have been cut away to let in the only two "fertilisers" needed—air and sun—the dimensions of the first pool or "reach" in the brook garden are decided upon. This must depend partly on the size and flow of the stream. If it is a chalk spring, from six feet to six yards wide, its flow will probably be constant throughout the year, for it is fed from the reservoirs in the heart of the hills. Then it needs little care except to clear its course, and the planting of its banks with flowers and stocking of its waters with lilies, arums, irises, and trout is begun at once. But most streams are full in winter and low in summer. On these the brook gardener must take a lesson from the beavers, and make a succession of delightful little dams, cascades, and pools, to keep his water at the right level throughout the year. Where there is a considerable brook these dams may be carried away in winter and ruin the garden. Stone or concrete outfalls are costly, and often give way, undermined by the floods. But there is a form of overflow which gives an added sparkle even to the waterfall, and costs little. Each little dam is roofed with thin split oak, overlapping like the laths of a Venetian blind when closed. This forms the bottom of the "shoot," and carries the water clear of the dam into the stream below. As the water runs over the overlapping laths it forms a ripple above each ridge, and from the everlasting throb of these pleats of running water the sunlight flashes as if from a moving river of diamonds. Beside these cascades, and only two inches higher than their level, are cut "floodoverflows" paved with turf, to let off the swollen waters in autumn rains. With the cutting out of undergrowth and the admission of light the rank vegetation of the banks changes to sweet grass, clovers, woodruffe, and daisies, and the flowers natural to the soil can be planted or will often spring up by themselves. In spring the banks should be set thick with violets, primroses, and the lovely bronze, crimson, and purple polyanthuses. Periwinkle, daffodils, crocuses, and scarlet or yellow tulips will all flourish and blossom before the grass grows too high or hides their flowers. For later in the year taller plants, which can rise, as all summer wood-plants do, above the level of the grasses, must be set on the banks. Clumps of everlasting peas, masses of phloxes, hollyhocks, and, far later in the year, scarlet tritomas (red-hot pokers) look splendid among the deep greens of the summer grass and beneath the canopy of trees. For it must be remembered that the brookside garden is in nearly every case a shaded garden, beneath the tall trees natural to such places. All beautiful flowering shrubs and trees, such as the guelder-rose, the pink may, the hardy azaleas, and certain of the more beautiful rhododendrons will aid the background of the brook garden, and flourish naturally in its sheltered hollow. There is one "new" rhododendron, which the writer saw recently in such a situation, but of which he does not recollect the name, which has masses of wax-like, pale sulphur flowers, which are mirrored in a miniature pool set almost at its foot. This half-wild flower garden pertains mainly to the banks of the brook gully, and not to the banks of the brook itself. It is in the latter, by the waterside, that the special charm of these gardens should be found. It is the nature of such places to have a strip of level ground opposite to each of the curves of the stream. All the narcissi, or chalice-flowers, naturally love the banks of brooks—

"Those springs
On chaliced flowers that lies."

These will grow in great tufts and ever-increasing masses, multiplying their bulbs till they touch the water's edge. Not only the old pheasant's-eye narcissus, but all the modern and splendid varieties in gold, cream, white, and orange, grow best by the brookside. By these, but on the lower ground almost level with the water, big forget-me-nots, butterburs, and wild snake's-head lilies should be set, and all the crimson and white varieties of garden daisy. Lily-of-the-valley, despite its name, likes more sun than our brook garden admits except in certain places; but certain of the lilies which flourish in the garden beds grow with an added and more languid grace on the green bank of our flower-bordered brook, and the American swamp-lily finds its natural place. Then special pools will be formed for the growth of those plants, foreign and English, which love to have their roots in water-soaked mud or the beds of running streams, while leaves and flowers rise far above into the light. Other pools should become "beds" for the water-flowers that float upon the surface. In the slang of the rock garden the plants living and flourishing on upright rocks are called "verticals." If we must have a slang for the flora of the brook garden we will term them "horizontals"— the plants that lie flat on the water surface, and only use their stems as cables to anchor them to the bottom of the stream. Of these we may plant, in addition to the white water-lily and the yellow, the crimson scented water-lily and the wild water-villarsia. White water-crowfoot, water-soldier, and arrowheads will form the fringe of the pool. But the crowning floral honour of the brook garden is in the irises set in and beside its waters, chief among which are the glorious irises of Japan— purple, blue, rose-colour, and crimson—the pink English flowering rush, big white mocassin flowers, New Zealand flax, and pink buckbean, and bog arum. The great white arum of the greenhouse is quite hardy out of doors if it is planted eighteen inches below water, and blossoms in the brook.

[Illustration: WATER VIOLET AND WILD IRIS. *From photographs by E. Seeley.*]

The brook garden is like a colony. It is always extending its range, following the course of the stream. Each year adds a little more to the completeness of the lower pools, and each year some yards of the upper waters and their banks are brought into partial harmony with the lower reaches. In one perfect example of this kind of garden, under the Berkshire downs, the succession of trout-pools, water gardening, half-wild banks, and turf-walk stretches for nearly a mile among the fields in a narrow glen, unseen from either side, except for its narrow riband of tree-tops among the fields; but within its narrow limits it is glorious with flowers, cascades, pools full of trout, set with water-plants in blossom, and the haunt of innumerable birds. Even the wild ducks ascend to the topmost pools, and are constantly in flight down the narrow winding vistas of grass, water, and trees, which they, like the kingfishers and water-hens, seem to think are set out for their especial pleasure.

COTTAGES AND CAMPING OUT

This is supposed to be a "business" country, but one wonders why new wants which accompany any change of daily habit are so slowly realised. Take, for instance, the annual migration to the Thames Valley, which has assumed proportions never reached before. Beyond the enlargement of the riverside inns, little has been done to meet this new taste of English families for rustic life in place of the seaside; and though the thousands of visitors to the "happy valley" of our largest river do contrive to enjoy a maximum of fresh air and outdoor life, this is often accompanied by a needless sacrifice of comfort. If any improvements in the conditions of life by the river can be suggested and put into practice, these will certainly benefit other districts. The profits accruing to intelligent provision for such a demand should also be considerable. But the first condition is that the wants and wishes of those who take their pleasure in this way should be properly understood.

The boating part of the river life is quite well organised; indeed, it would be difficult to improve upon it. Its convenience and elasticity is remarkable. The way in which the leading boatbuilders provide craft of all descriptions, which may be left by their hirers at any point on the river, to be brought back to Oxford or Reading by train, is beyond all praise. It is a triumph of good sense and management. But boating is only part of the amusement of the holiday, just as bathing is at the seaside. The real object with which an evergrowing number of visitors have adopted the river life is in order to spend the utmost length of time out of doors and in beautiful scenery. To this end they need accommodation of a special kind. The large hotel, with its inducements to spend much time over meals and indoors, is wholly out of place for such a purpose. What is needed is a cottage which can be rented either wholly or in part, or actual camp life under tents. The latter is now not confined to boating-men travelling up or down the river. It is enjoyed partly as an annexe to up-river houseboats; more often as "camping out" for its own sake, the tents being pitched near the river, but in complete detachment from any other habitation, fixed or floating. In these tents whole families of the well-to-do classes now elect to live, sometimes for weeks; rising early, bathing in the river, sometimes cooking their own food, or more often employing a servant or local man-of-all-work to do this, taking their meals in the open, and using the tents only to sleep in, or as a shelter from rain. Even little children now share the delights of this *al fresco* life, which realises their wildest dreams of adventure, and is by general consent as wholesome as it is entrancing. Whether their elders derive as much pleasure as they might from the same environment is doubtful. The business is not properly organised, and only half understood by the greater number of those who are nevertheless so well pleased by the experiment that they are anxious to repeat it. Sporadic camping out involves too much fetching and carrying. Tradesmen do not "call" at isolated tents in a riverside meadow, and all commodities have to be fetched by the campers. On the other hand, sociable camping out, when several groups set up their tents in proximity, needs proper arrangement. Philosophers may see in it the evolution of the social life from its primitive elements, with the growth of division of labour and reciprocal good offices. English families would usually prefer the sporadic tent, if it were not for the hard work involved. But if camping out is to be a real success, such understandings and arrangements must be made. Where this is not done the result is a failure, obvious to the passer-by. Separate and unsightly fires for cooking, and untidiness, because there are no "hours" for performing the light but necessary domestic work, are common objects of individualism on the camping ground. Yachts, which are self-maintaining, never have clothes hanging in the rigging after 8 a.m. when in harbour, and the self-respecting camp must not fall behind this example.

The camp in the country should have its communal kitchen in a wooden movable house, in which meals can be cooked, and from which it should be possible to purchase food as required. Here is an opening for commercial enterprise. The tourist agencies might rent camping grounds and supply tents on hire, with kitchens and all proper necessaries for living under canvas. They do this with great success for travellers in the East, and at a moderate cost. In England tents, if not so luxurious as those provided from Egypt for life in Palestine, are very cheap, and need no transport animals. But such a firm could easily make them removable by arranging for them to be called for and taken up river a few stages, as the boats are. The hire could be fixed at so much per tent, and a camp servant could also be provided. Commissionaires and ex-soldiers with good characters could be found employment in the early autumn, when they now find it difficult to earn a wage. They thoroughly understand not only the management of tents, but the duties of a camp. Rain-proof tents with movable board floors would be provided from London in uncertain weather on the receipt of a wire, for life under canvas is quite pleasant even if the hours are not all serene, if the interior is kept dry.

Though a new departure in this country camping out is part of the ordinary and well-understood amusements of the eastern cities of the United States. The whole State of Maine is practically a State reserve for this, the most popular form of holiday-making in America. Its forests, rivers, and lakes are one vast playground and public sporting domain, which is enjoyed almost entirely by means of camping out and boating. The rivers teem with State-reared trout, of which as many are allowed to be caught as can possibly be consumed by the party. The woods are free to shoot in, with a limit for deer and caribou; State-provided guides are employed at a fixed wage. At regular intervals along the rivers are the camping grounds, each under the control of a camp agent, who arranges for the comfort and convenience of the travelling host of tent-dwellers. Each "base" is properly organised and supplied, and visitors can purchase necessaries, in addition to the fish and birds which fall to rod and gun. Ladies and children are among those who enjoy the pastime most keenly, amusing themselves by the river and among the woods while the husbands hunt or fish.

The "residential cottage" is perhaps the safer basis for the complete outdoor life, though it tends to reduce the number of hours spent in the open. Habit is too strong when once we are under a roof. It is evidence of the habitable nature of many of our much-abused cottages that in the Thames-side villages a great proportion are now occupied for several months in the year by people who, though willing to pay for simple accommodation, will not tolerate dirt, squalor, or want of sanitation. To their surprise they have found hundreds of cottages, homely, but not uncomfortable, kept with scrupulous neatness, and furnished by no means badly. Nearly all have ample kitchen accommodation, fair beds, and an equipment of glass, china, and crockery, which shows how cheap and good are the necessaries of life in England. The well-to-do agricultural labourer and his wife, whose children are out in the world, the village artisans, small tradesfolk, and "retired" couples are the owners or occupiers, and now let their rooms at from £1 to £1 10s. per week, from June till the middle of September. The results are good in every way. Visitors are pleased at what seems a cheap holiday, and the letters of the rooms save money for the winter, and realise in a pleasant way that their later years have fallen on good times. It is also an encouragement to landowners to build good and picturesque cottages. For the first time they see their way to charging a fair rent on their outlay. The town comes to help the country, and the country sees in the movement a hopeful future.

NETTING STAGS IN RICHMOND PARK

About the opening of the year I went to see the big stags netted in Richmond Park for transfer to Windsor. Last season this unique and ancient hunting had to be put off till February. There was too much "bone" in the ground to make riding safe. When the frost gave, the stags were more than usually cunning, and were helped by more than their usual share of luck. One fine stag charged the toils at best pace, and, happening to hit a rotten net, burst through, and went off shaking his antlers as proudly as if he had upset a rival in a charge. Another took to the lake, and after playing Robinson Crusoe on the island for some time, swam across to the wood, took a standing leap out of the shallow water on the brink over the paling, and laid up in Penn Wood.

It was on a lovely mellow January morning, after just a touch of frost, with haze and mist veiling the distant woods, a winter sun struggling to make itself seen, and all the birds, from the mallards on the lakes to the jackdaws in the old oaks, beginning to talk, but with their minds not quite made up as to whether they should take a morning flight or stop where they were, when the business of setting up the toils began.

This, which is probably managed in exactly the same way as when Queen Dido arranged to give a day's sport to good Aeneas, is carried out according to the ancient and unvarying tradition of this royal and ancient park. Nor were we allowed to forget that in this case, too, the stags were being taken by the servants of a queen. Everything was ready for the transport of the stags to Windsor, and in the foreground was a good strong wooden cart, painted red and blue, and inscribed in the largest capitals with the words, "Her Majesty's cart."

The art and practice of taking the stags in the toils is carried out in this wise. A body of mounted men, under the orders of the superintendent of the park, ride out to find the herds of red deer. They then ride in and "cut" out the finest stags, and, spreading out in a broad line, chase them at the utmost speed of horse towards that quarter of the park where the nets are spread. Some two hundred yards in front of the nets two deerhounds are held, and slipped as the stag gallops past—not to injure or distress him, but to hurry him up and distract his attention from the long lines of nets in front.

The stags were known to be full of running, and resourceful; consequently the number of riders who had been asked to help was rather larger than usual. Even so they had to make a wide sweep of the Southern Park before they found their deer, and had a racing burst of more than a mile and a half before they brought them round. Meantime, while they are away on their quest, let us inspect the ancient contrivance of the toils. They are heavy nets of rope, thick as a finger, and with meshes not more than ten inches square—very strong, and to our eyes almost too solid and visible. Partly to render them less conspicuous, the line—at least one hundred yards long—is set in a long, narrow depression or shallow drain, running from a wood on the Richmond side of Penn Pond down to a small pool. Just in the centre of this line is a most ancient pollard oak, the crown of which will hold eight men easily, ready to spring down to earth and seize the deer as the nets fall on him. In this most appropriate watch-tower the keeper in command at the toils, and several of his helpers, ensconced themselves. The Richmond stags, though so constantly in the sight of the crowds of visitors to the park, are among the boldest and gamest of all park stags. One, who was more especially the object of the day's chase, jumped a paling 6 ft. 3 in. high the day before, merely for amusement. Those sometimes transferred to the paddocks at Ascot for hunting with the Royal Buckhounds were noted for their courage and straight running. Perhaps the most famous was old Volunteer, whose latest exploit was to give a run of nearly thirty miles, at the end of which he was not taken. Having had his day out, and not being taken up in the cart as usual, he made his way home by night, jumped into his paddock, and was found there next morning!

Holloaing, long and loud, was now heard from the east. Keen was the keeper's glance as he looked, not to the sound, but along his line of nets, the top at least eight feet from the ground, lightly hitched on thick saplings, while an ample fold of some four feet more lay upon the ground. Before and behind, the dead and tangled bracken broke the line; the props were of natural wood, and the tawny nets themselves made no break in the general colour of the hillside. Then the shouting came louder down the wind. Where were they? Not coming "up the straight" certainly, for no stags were visible and the hounds were not slipped. Suddenly from above us three big red stags came galloping obliquely down the hill, not as they are represented in pictures with muzzles up and horns back, but at high speed for all that; and though they carried their horns erect, their sides were heaving and the smoke coming out of their nostrils. They saw the nets, but determined to push through them. One charged them gallantly head first, and as the thick meshes fell tumultuously over his head and back, the second jumped the falling toils twenty yards to his left, taking them most gracefully, as if he were doing a circus trick. Down from the tree sprang the keeper and his men, and seized the helpless stag, while the second, which had jumped and won, stood panting and looking over his shoulder to see what curious game this was. The third broke back and disappeared.

Perhaps the most strange thing was the calm self-possession of the netted stag. The astonishing catching power of a net held him enmeshed at all points. His muzzle was held by one mesh, his horns by three or four; all four feet were caught also. In addition, about eight men kindly caught hold of his horns, legs, and back, to prevent him hurting himself. This he was far too clever to do. He just lay quiet, calmly regarding the fun with his upper eye, and wondering when the deuce they were going to take him "out of that." In a very few minutes his feet were buckled together by soft straps, and a saw trimmed off his antler tops, for which we felt sorry, but there was not room for them in the "compartment" he was to travel in. It is only when a stag lies close before you on the ground that you realise that he is not a "slab-sided," flat-ribbed animal, but a bulky, well-rounded beast. It took six men to lift him on to the bed of fern in "Her Majesty's cart," and when there he quickly twisted round, and lay couched, bound but not subdued, calmly regarding the scene over the side of his cart. A nice lot of chopped mangold root had been put in his box, and we hope he enjoyed his lunch in the train on his way to Windsor.

[Illustration: A NETTED STAG. *From a drawing by Lancelot Speed.*]

The next drive was far more rapid, and its results more exciting. The stags were again brought round from above Penn Pond, then through the oak grove below White Lodge, and came galloping up the long side of the slope, straight for the nets. Then the brace of deerhounds, which, like the keeper, seemed to know the game thoroughly, were slipped, and most beautiful they looked, one laying out, lithe and low, just parallel with the haunch of one stag, the other driving the brace below. The single stag charged the nets and was enveloped as before, but the other brace broke back and escaped.

Four in all were taken during the day, without accident or mishap. One of the keepers did have an accident of a rather curious kind, when assisting to catch stags at Buckhurst Park in Kent. He was galloping as hard as he could, driving a stag, when his horse cannoned up against another deer which was lying crouched in the fern, as deer sometimes do. The horse went a complete somersault, and its rider was badly bruised and hurt, though no bones were broken.

RICHMOND OLD DEER PARK

If Henry VII.'s palace at Richmond still stood by the riverside, we should have a second Hampton Court at half the distance from London. It was almost the first of the fine Tudor palaces in this country, built very stately, with a prodigious number of towers, turrets, cupolas, and gilded vanes, on a site as fine as that of Wolsey's competing pile higher up the river. But though the palace has gone, the park is left. It is the precinct now called the Old Deer Park, in which not one in ten thousand of those who visit and enjoy the park on the hill which we call Richmond Park has ever set foot, except in the corner furthest from the river to see a horse-show or a cricket-match. Old it certainly is. The park on the hill, venerable as it looks now, is only a thing of yesterday in comparison with it. Charles I. made the latter, and the Penn Ponds were dug by the

Princess Amelia. The Old Deer Park was a Royal demesne when the Saxon Kings had their palace at Sheen, before it was given its new name of Richmond by the first Tudor, after the Castle in Yorkshire from which he took his title when a subject. In the middle of this ancient and forgotten park, forgotten because it is neither reserved for the pleasure of the Sovereign nor thrown open for the enjoyment of his subjects, it was lately proposed to build a scientific laboratory, to supplement the work of the observatory, which is mainly employed in magnetic observations and in testing thermometers and chronometers. The proposal is an instance of the mischief which may be done by precedent, and of the way in which Royal favour may be misused quite unconsciously by persons who forget that the circumstances which lent grace and propriety to a concession at one time may be so altered later that to presume on it is an error of judgment. George III. instructed Chambers, the architect, who had been doing work for him at Kew, to erect an observatory in the Old Park. It was a thoughtful act, at a time when there were no public funds for the encouragement of science, and when the study of astronomy was still regarded partly as something peculiarly under Royal patronage because its practical use was to keep and make records to ensure the safe navigation of his Majesty's ships.

The application to erect new buildings was refused, for a place like the Old Deer Park, if kept open and wild, and not built upon, has a present and future value to the health and happiness of millions of people beyond any calculation or power of words.

It does not need much imagination to make this forecast. But as few people have ever made what, in the old words of forest law, was called a "perambulation" of the park, some description of its present condition and appearance may help to form an opinion. It is the largest and finest riverside park in England. It covers nearly four hundred acres, but this great area, as large as Hyde Park, is shaped and placed so as to gain the maximum of beauty and convenience from its surroundings. On the London side it has for neighbour the whole depth of Kew Gardens, from the road at the back to the river at the front—two hundred and eighty acres of garden and wood. But whoever first acquired the land for the park, whether Norman or Saxon, very rightly thought that the feature to be desired was to make the most of the river-front, where the

Thames, pushing into Middlesex, cuts "a huge half-moon, a monstrous cantle, out." Whether by accident or design, the park is like a half-open fan, narrowest at the back, which is the ugly or plain side, near the road, and with its widest part unbosoming on the Thames. From back to front it is some half-mile deep; but the Thames front extends for a mile along one of the most beautiful river scenes in England.

On the Kew Gardens border it lies against what was, until a few years ago, the wild and private part of Kew. To this it served as an open park, where all the birds drew out to sun themselves and feed. So they do still. Along the margin are scattered old beech trees, and a wilderness of long grass and flowers, where wood-pigeons, thrushes, pheasants, crows, jays, and all the smaller birds of the gardens may be seen sunning themselves. The narrow end or "stick" of the "fan," near the road, is leased to a cricket club, and cut off from the greater area by a belt of young plantation. In this a brood of partridges hatches nearly every year, though what becomes of the birds later is only conjectured. Beyond this cross-belt the whole area of the park stretches out, ever widening, and with an imperceptible fall, to the Thames. It is studded here and there with very large and very ancient trees, and is one of the largest and least broken areas of ancient pasture, whether for deer or cattle, in England. Until lately the old observatory was the only building upon it, and the turf was unbroken. But recent years have added two disfigurements. One is a large red building with skylights, connected with the games and athletic sports, which have found a more or less permanent home in the upper part of the park, where the annual horse-shows are held, uses for which that part of the ground is well suited. The other is a permanent and very deplorable blemish, made purposely, in the interests of the popular game of the hour. The greater part of this fine park has been leased to a private golf club. Golf, as every one knows, originally flourished on sand dunes, which are about as completely the natural opposite of an old flat park of ancient pasture as can be found in this country. The golf club have been allowed to do what they can to remedy this defect of Nature by converting the Old Park into a sand dune, and this they have done by digging holes and throwing up dozens, or scores, of bunkers. But the margins of the park are quite unspoilt, and the river-front is the wildest and the freest piece of Nature left near London. It is completely bounded by an ancient moat, beyond which lies the towing-path, and beyond that the river and the ancient and picturesque front of Isleworth. The path between the moat and the river is set with ancient trees, mostly horse-chestnuts and beech, in continuous line. Under their branches and between their stems the visitor in the park sees a series of pictures, framed by trees and branches, of the Queen Anne houses and rose-gardens of Isleworth, the old church with its tower and huge sun-dial, the ferry and the old inn of the "London Apprentice," the poplars and willows of the Isleworth eyots, the granaries and mills where the little Hounslow stream falls in, and further Twickenham way the gardens of the fine villas there, while towards London the pavilions and park of Syon House begin. At the present moment the margin of the Old Deer Park and its moat give a mile of beauty and refreshment. No one has troubled to mow the grass or cut the weeds, or clear the moat, or meddle with the hedge beyond it. So the moat, which is filled from the river when necessary, and is not stagnant, is full of water-flowers, and quite clear, and fringed with a deep bed of reeds and sedges. In it are shoals of dace, and minnow, and gudgeon, and sticklebacks, and plenty of small pike basking in the sun. The largest and bluest forget-me-nots, and water-mints, and big water-docks and burdocks flourish in the water, and the hedge beyond is full of sweet elder in flower, and covered with wild hops. Huge elms, partly decaying, and a dark grove of tall beeches line the park near the moat, and besides water and flowers there is shade and the motion of leaves. If the proposal to build on such a site leads to a better knowledge of what this ancient park really is, and its value to the amenities of the capital, it will have done good, not harm. The late Queen recently presented the cottage in the reserved part of Kew Gardens and its precincts for the use of the public. It would seem that a similar sacrifice has been made by Royalty in the case of the Old Deer Park, but that the public are excluded by the Office of Woods and Forests, which has charge of it, and the park neglected and disfigured. If it were put on the same footing as Richmond Park upon the hill, and communication were open between the park and Kew Gardens at proper hours, an unequalled domain, still the property of the Crown, but enjoyed within reasonable limitations by every subject, would be open from Kew Green practically to Kingston. The line from the boundary of the Old Deer Park is taken on by Richmond Green, and the towing-path to the Terrace Gardens, formerly the property of the Duke of Buccleuch, and now of the Richmond Corporation, thence by the terrace and the open slope under it to Richmond Park, through Sudbrook Park to Ham Common, a series of varied scenery unrivalled even in the valley of the Thames.

FISH IN THE LONDON RIVER

The capture of a 4-lb. grilse in the Thames estuary in December, 1901, raised some hopes that we might in course of time see salmon at London Bridge. Mr. R. Marston, a great authority, in an article on "The Thames a Salmon River," in the *Nineteenth Century*, has given many reasons why he fears that this will not be realised. The question is not so much whether the salmon can come up, as whether the smolts, or young salmon, could get down through the polluted water. But the experiments made are interesting and deserve every encouragement, and it may be hoped that money will be forthcoming to make more.

As regards other fish than salmon, their return has been going on steadily since 1890; and their advance has covered a distance of some twenty miles—from Gravesend to Teddington. The first evidence was the reappearance of whitebait, small crabs, and jelly-fish at Gravesend in 1892. In 1893 the whitebait fishermen and shrimp-boats were busy ten miles higher than they had been seen at work for many years. The condenser tubes of torpedo-boats running their trials down the river were found to be choked with "bait," and buckets of the fish were shown at the offices of the London County Council in Spring Gardens. It was claimed that this evidence of the increased purity of the water was mainly due to the efforts of the Main Drainage Committee of the London County Council. There is abundant evidence that this claim was correct, for instead of allowing the whole of the London sewage to fall into the Thames at Barking and Crossness, the County Council used a process to separate all the solid matter, and carried it out to sea. The results of the first year's efforts were that over two million tons were shipped beyond the Nore, ten thousand tons of floating refuse were cleared away, and the liquid effluent was largely purified. It was predicted at the time that if this process was continued on the same scale it would not be long before the commoner estuary fishes appeared above London Bridge, even if the migratory salmon and sea-trout still held aloof. Unfortunately there has been some deviation from the methods of dealing with the sewage, a change from which we believe that some of the officials concerned with the early improvements very strongly dissented, that has to some extent retarded the advance of the fish. But in 1895 a sudden "spurt" took place in their return. Whitebait became so plentiful that during the whole of the winter and spring the results were obvious, not only to naturalists, but on the London market. Whitebait shoals swarmed in the Lower Thames and the Medway, and became a cheap luxury even in February and March. They were even so suicidally reckless as to appear off Greenwich. Supplies of fresh fish came into the market twice daily, and were sold retail at sixpence per quart. The Thames flounders once more reappeared off their old haunt at the head of the Bishop of London's fishery near Chiswick Eyot. Only one good catch was made, and none have been taken since; but this had not been done for twelve years, and there is a prospect of their increase, for, in the words of old Robert Binnell, Water Bailiff of the City of London in 1757, we may "venture to affirm that there is no river in all Europe that is a better nourisher of its fish, and a more speedy breeder, particularly of the flounder, than is the Thames." Eels were also taken in considerable numbers between Hammersmith and Kew; but the main supply of London eels came from Holland even in the days of London salmon. In a very old print of the City, with traitors' heads by the dozen on London Bridge, "Eale Schippes," exactly like the Dutch boats lying at this moment off Billingsgate, are shown anchored in the river. Besides the estuary fish which naturally come *up* river, dace and roach began to come *down* into the tideway, and during the whole summer the lively little bleak swarmed round Chiswick Eyot. Later in the year the roach and dace were seen off Westminster, and several were caught below London Bridge, and in 1900 roach were seen and caught at Woolwich, but were soon poisoned and died. In August the delicate smelts suddenly reappeared at Putney, where they had not been seen in any number for many years. Later, in September, another migration of smelts passed right up the river. Many were caught at Isleworth and Kew, and finally they penetrated to the limit of the tideway at Teddington, and good baskets were made at Teddington Lock.

[Illustration: BREAM AND ROACH. *From a photograph by E. Seeley.*]

This additional evidence of the satisfaction of the fish with the County Council does not quite satisfy us that the London river is yet clean enough to give passage to the migratory salmon. It is encouraging to the County Council, who deserve all the credit they can get; but there is little doubt that the best evidence that the river is fit for the salmon would be the spontaneous appearance of the salmon themselves.

Since the middle of June, 1890, large shoals of dace, bleak, roach, and small fry have appeared in all the reaches, from Putney upwards. A few years ago hardly any fish were to be seen below Kew during the summer, and these were sickly and diseased. Last year they were in fine condition, and dace eagerly took the fly even on the lower reaches. Every flood-tide hundreds of "rises" of dace, bleak, and roach were seen as the tide began to flow, or rather as the sea-water below pushed the land-water before it up the river. At high water little creeks, draw-docks, and boat-landings were crowded with healthy, hungry fish, and old riverside anglers, whose rods had been put away for years, caught them by dozens with the fly. Sixty dozen dace were taken, mainly with the fly, in a single creek, which for some years has produced little in the way of living creatures but waterside rats. I counted twenty-two "rises" in a minute in a length of twenty yards inside the eyot at Chiswick. During one high tide in July a sight commonly seen in a summer flood on the Isis or Cherwell was witnessed not sixty yards from the boundary stone of the county of London. The tide rose so far as to fringe several lawns by the river with a yard or two of shallow water, and the fish at once left the river and crowded into this shallow overflow, their backs occasionally showing above it, to escape the muddy clouds in the tidal water. There were hundreds of fish in the shoals, of all kinds and sizes, from dace nine inches long, with a few roach, to sticklebacks. These fish are probably the descendants of spawn laid in the *tidal* parts of the river, on the gravel-beds and weeds. Doubtless the quantity of fresh water from the spring rains contributed something to the result, but the spawn must have hatched far more successfully than usual.

[Illustration: A GRAMPUS AT CHISWICK. *From a drawing by Lancelot Speed.*]

Rivermen on the tidal Thames always distinguish between eels and "fish." The former are also increasing greatly. The sole survivor of the "Peter boats" left on the river is saved from disappearing like the rest of the race by eel-fishing. Formerly these boats, whose owners lived and slept on board them for six months in the year, were quite successful in catching eels and flounders. In the Chiswick parish registers a number of those married or buried are entered as being "fishermen," which clearly means that that was their business in life. The number of professed eel catchers' boats gradually dwindled to one, and the owner of this catches a fair quantity of most excellent eels, those taken off Mortlake, opposite the finish of the University boat-race, being especially fine in flavour. Another eel-like fish, formerly taken in great numbers, and of the finest quality, but now almost forgotten, is also returning. This is the lampern. Lamperns, unlike eels, come into the rivers to spawn, and go back to the sea later or to the brackish waters. Men employed in scooping gravel out of the river at Hammersmith, lately noticed numbers of lamperns coming up on to the gravel-beds at low-water, and moving the gravel into little hollows, previously to dropping their spawn. Twelve years ago the great body of the migrating lamperns were all poisoned by the river, and lay in tens of thousands in the mud at Blackwall Point. As they have now succeeded in getting up to spawn, the shoals may be seen next year in something like their old numbers. The flounders have not yet reappeared to stay. Porpoises come up above London nearly every year. The first I saw were two above Hammersmith Bridge early on that momentous May morning in 1886, when Mr. Gladstone's first Home Rule Bill was thrown out. I had been up with a friend to hear the result of the division, and had seen the wild joy which followed its announcement in the lobby, and then walked home at dawn, and so met the early porpoises. A few years later a fine grampus was found one night lying half dead by the bows of one of the torpedo-boat destroyers at Chiswick. Its "lines" struck the expert minds there as so good that it was carefully measured, and the results were found to correspond almost exactly with a mathematical curve—I think called a curve of sines. The hollow over the blow-hole was filled up with mud and measured over, and here there was a little discrepancy. The mud was removed, and the measurement taken over the surface of the hollow, and the figures found to be what were expected.

CHISWICK EYOT

It has been said that Thames eyots always seem to have been put in place by a landscape gardener. Chiswick Eyot is no exception to the rule. It covers nearly four acres of ground, and lies like a long ship, parallel with the ancient terrace of Chiswick Mall, from which it is separated by a deep, narrow stream, haunted by river-birds, and once a famous fishery.

A salmon, perhaps the last, was caught between the eyot and Putney in 1812, though the rent of the fishery used to be paid in salmon, when it was worked by the good Cavalier merchant, Sir Nicholas Crispe. The close-time for the fishery was observed regularly at the beginning of the century, the fishing commencing on January 1st, and ending on September 4th. There are those who believe that with the increased purification of the Thames, the next generation may perhaps throw a salmon-fly from Chiswick Eyot. In the early summer of 1895 a fine porpoise appeared above the island. At half-past eight it followed the ebb down the river, having "proved" the stream for forty miles from its mouth, and being apparently well pleased with its condition. At Putney it lingered, as might be expected of a Thames porpoise, opposite a public-house. Two sportsmen went out in a boat to shoot it; instead, they hit some spectators on the bank. Flowers abound on the eyot. The irises have all been taken, but what was the lowest clump, opposite Syon House, has lost its pride of place, for now there are some by the Grove Park Estate below Kew Bridge. The centre of the eyot is yellow with patches of marsh-marigold in the hot spring days. Besides the marsh-marigolds there are masses of yellow camomile, comfrey, ragged robin, and tall yellow ranunculus, growing on the muddy banks and on the sides of the little creeks among the willows, and a vast number of composite flowers of which I do not know the names. Common reeds are also increasing there, with big water-docks, and on the edge of the cam-shedding of the lawn which fronts my house some of the tallest giant hemlocks which I have ever seen, have suddenly appeared. I notice that in Papworth's views of London, published in 1816, arrowhead is seen growing at the foot of the Duke of Buckingham's water-gate, which is now embedded at the back of the embankment gardens at Charing Cross. There is still plenty of it opposite Hammersmith Mall, half a mile below Chiswick Eyot. The reach opposite and including the eyot is the sole piece of the natural London river which remains interesting, and largely unspoilt. I trust that if urban improvers ever want to embank the "Mall" or the eyot, public opinion will see its way to keeping this unique bit of the London river as it is. Already there have been proposals for a tram-line running all the length of the Mall, either at the front or behind it. The island belongs to the Ecclesiastical Commissioners. There is a certain sense of the country about the eyot, because it is rated as agricultural land, though its lower end is inside the London boundary. The agriculture pursued on it is the growing of osiers. These, frequently inundated by high tides, and left dry when the ebb begins, are some of the finest on the Thames. At the present moment (January 5, 1902) they are being cut and stacked in bundles. In the spring the grass grows almost as fast between the stumps as do the willow shoots. This is cut by men who make it part of the year's business to sell to the owners of the small dealers' carts and to costers. Formerly, when cows were kept in London, it was cut for their use. During the year of the Great Exhibition milk was very scarce, and this grass, which was excellent for the stable-fed cows, fetched great prices. In the summer the willows, full of leaf, and exactly appropriate to the flat lacustrine outline of the eyot and the reach, are full of birds, though the reed-warbler does not always return. He was absent last year. He is locally supposed to begin his song with the words "Chiswick Eyot! Chiswick Eyot!" which indeed he does pretty exactly. Early on summer mornings I always see cuckoos hunting for a place to drop an egg. In the summer of 1900 a young cuckoo was hatched from a sedge-warbler's nest, and spent the rest of the summer in the gardens opposite this and the next houses. All day long it wheezed and grumbled, and the little birds fed it. In the evenings it used to practise flying, and at last flew off for good.

CHISWICK FISHERMEN

"Please, sir, a man wants to know if he can see you, and he has brought a very large fish," was the message given me one very hot evening at the end of July, at the hour which the poet describes as being "about the flitting of the bats," plenty of which were just visible hawking over the willows on the eyot. Thinking that it was an odd time for a visit from a fishmonger, I was just wondering what could be the reason for such a request when I remembered a talk I had had at the ferry a week or two before on the subject of the continued increase of fish in the London Thames. It turned out to be as I expected; my visitor was one of the last local fishermen, and brought with him a splendid silver eel, weighing nearly 4 lb., taken in his nets that evening just opposite Chiswick Eyot. It was the largest eel taken so low down for some years, and when held up at arm's length, was a good imitation of one of Madame Paula's pythons in the advertisement. He was anxious that I should come out for an evening's netting and see for myself how clear the water now is, and how good the fish. The previous summer, about the same date, I had asked him to see what he could catch in an evening as specimens; he had returned with over ninety fish, dace, roach, eels, barbel, and smelts, many of which were exhibited alive the next day before a good many people interested in the purification of the Thames. As a further proof I forwarded the big eel to the previous chairman of the London County Council, under whose sceptre the marked improvement in the river began first to be felt, and begged his acceptance of it as a tribute from the river. Then I arranged to be at the old ferry next day at 6.30 p.m.

It was the end of a blazing hot London day when I went down the hard to the water's edge, among the small, pink-legged boys, paddling, and the usual group of contemplative workmen, who smoke their pipes by the landing place. The river was half empty, and emptying itself still more as the ebb ran down. The haze of heat and twilight blurred shapes and colours, but the fine old houses of the historic "Mall," the tower of the church, and the tall elms and taller chimneys of the breweries, which divide with torpedo boats the credit of being the staple industries of Chiswick, stood out all black against the evening sky; the clashing of the rivetters had ceased in the shipyard, but the river was cheerfully noisy; many eights were practising between the island and the Surrey bank, coaches were shouting at them, a tug was taking a couple of deal-loaded barges to a woodwharf with much puffing and whistling, and bathers, sheltered by the eyot willows, were keeping up loud and breathless conversations. "Not exactly the kind of surroundings the fishermen seeks," you will say; but, apparently, London fish get used to noise. Our boat was what I, speaking unprofessionally, should call a small sea-boat, but I believe she was built years ago at Strand-on-the-Green, the pretty old village with maltings and poplar trees that fringes the river below Kew Bridge. She was painted black and red, and furnished with a shelf, rimmed with an inch-high moulding inboard and drained by holes, to catch the drip from the net as it was hauled in. We were at work in two minutes. The net was fastened at one end to two buoys; these dropped down with the ebb, and formed a fixed, yet floating, point—if that is not a bull—from which the boat was rowed in a circle while one of the brothers who own the boat payed out the net. Thus we kept rowing in circles, alternately dropping and hauling in the net, as we slipped down what was once the Bishop of London's Fishery towards Fulham. There are still no flounders on the famous Bishop's Muds, but other fish were in evidence at once. Though the heat had made them go to the bottom, we had one or two at every haul. The two fishermen were fine specimens of strong, well-built Englishmen. The pace at which they hauled in the net, or rowed the boat round, was great;

the rower could complete the circle—a wide one—in a minute, and the net was hauled in in less time, if the hauler chose to. Dace were our main catch—bright silvery fish, about three to the pound, for they do not run large in the tideway; but they were in perfect condition, and quite as good to eat, when cooked, as fresh herring. For some reason the Jews of London prefer these fresh-water fish; they eat them, not as the old Catholics did, on fasts, but for feasts. They will fetch 2d. each at the times of the Jews' holidays, so our fisherman told me, and find a ready sale at all times, though at low prices. Formerly the singularly bright scales were saved to make mother-of-pearl, or rather, to coat objects which were wished to resemble mother-of-pearl. After each haul the fish were dropped into a well in the middle of the boat. A few roach were taken, and an eel; but the most interesting part of the catch was the smelts. These sea-fish now ascend the Thames as they did before the river was polluted. We took about a dozen, some of very large size; they smelt exactly like freshly-sliced cucumber. I stayed for an hour, till the twilight was turning to dark, and the tugs' lights began to show. We had by then caught seventy fish, or rather more than one per minute; a hundred is a fair catch on a summer evening. In winter very large hauls are made; then the fish congregate in holes and corners. In summer they are all over the river. When the net happens to enclose one of these shelter holes, hundreds may be taken. Consequently the two fishermen work regularly all through the winter. Sometimes their net is like iron wire, frozen into stiff squares. In a recent hard winter the ice floated up and down the London Thames in lumps and floes; yet they managed to fish, and made a record catch of two thousand in one tide. I believe that if the Conservancy and the County Council go on as they are doing, we shall see the flounder back in the river above bridges, and that possibly sea-trout may adventure there too; though unless the latter can get up to spawn, there can be no regular run of sea-trout. But they probably also act like grey mullet, and run up the estuaries merely for a cruise.[1]

[Illustration: SMELTS. *From a photograph by E. Seeley.*]

The last of the "Peter-boat" men mentioned in a previous chapter, has other claims to notice than that of being the only survivor of an ancient outdoor industry. He has given evidence before more than one committee of the House of Commons on the state of the river and the condition of its waters, and is the oldest salesman in that curious survival of antiquity, the free eel market held at Blackfriars Stairs on Sunday mornings; and, in addition, he has added to his original industry another branch of "fishing" of a different kind, of which he is acknowledged to be the greatest living exponent. He is an expert at grappling and "creeping" for objects lying on the bed of the river, work for which his life-long acquaintance with the contours of the bottom and the tides and currents makes him particularly well fitted. Consequently he is now regularly employed by many firms and shipping companies to fish up anything dropped overboard, whether gear or cargo, which is heavy enough to sink. The oddest thing about this double business is that all the summer, while he lies and sleeps in his "Peter-boat" at Chiswick, he is in receipt of telegrams whenever an accident of this kind chances to happen, summoning him down river, to the Docks or the Pool, and these telegrams are delivered to him (I think by the ferryman) on his "Peter-boat." But the regular time for this other Thames "fishery" is in winter. Then the eels "bed," *i.e.*, bury themselves in the mud, and the eel man goes either "gravelling," that is, scooping up gravel from the bottom to deepen any part of the channel desired by the Conservancy, or doing these odd salvage jobs. Getting up sunken barges is one side of the business. These are raised by fastening two empty barges to them at low tide, when the flood raises all three together, owing to the increased buoyancy. But of "fishing" proper he has had plenty. He hooked and raised the steamship *Osprey's* propeller, which weighed six tons. This was done by getting first small chains and then large ones round it, and fastening them to a lighter. Half-ton anchors, casks of zinc, pigs of lead, copper tubes, ironwork, ship-building apparatus, and the like, are common "game" in this fishery. Other commodities are casks of pitch, cases of pickles, boxes of champagne, casks of sardines in tins, bales of wool, and even cases of machinery.

This form of Thames fishery increases rather than diminishes. Years ago he picked up under London Bridge a case of watches valued at £1,500. He was only paid for the "job," as the loss was known and it was not a chance find. Another and more sportsmanlike incident was an "angling competition," among himself and others in that line, for some cases of rings which a Jew, who became suddenly insane, threw into the river off a steamer. He caught one case, and another man grappled the other. Sometimes in fishing for one thing he catches another which has been in the water for months, as, for instance, a whole sack of tobacco, turned rotten. I do not know who "that young woman who kept company with a fishmonger" was, though he assumes that I do. But he certainly rescued her, and a gentleman who jumped off London Bridge, and several upset excursionists on various parts of the river. Also, as will be guessed, he has caught or picked up a good many corpses. I hear, though not from him, that on the Surrey side five shillings is paid for a body rescued, and on the Middlesex side only half-a-crown; so Surrey gets the credit of the greater number of the Thames dead. His life-saving services have been very considerable, though he does not make much account of them. He was instrumental in saving two women and six men on one occasion, and on another "three men and a soldier." The distinction is an odd one, but it holds good in the riverine mind.

[1] At the close of the season 1901-1902 in March, one of the men tells me that it has been the best year he has known. He caught sixteen eels one night with the net only. Very fine bream have also appeared as low as Hammersmith.

BIRDS ON THAMES RESERVOIRS

Now that every large town and many small ones are adding new reservoirs, often of great size, to hold their water supply, these artificial lakes play an important and increasing part in the wild life, not only of the country, but of cities, and even of London itself. Immense reservoirs have been made near Staines, and others are being added close to the London river. These quiet sheets of water, carefully protected from intrusion for fear of any pollution of the water, form artificial sanctuaries which not only fill with fish, which the water companies encourage, to eat the weeds and insects bred in the weeds, but attract wild-fowl of very many kinds in ever-increasing numbers. In Hertfordshire the artificial lakes near Tring made to supply the Grand Junction Canal are carefully preserved, and have a large and resident population of wild-fowl (we believe a bittern bred there recently, and the great crested grebe is common), and some of the new London reservoirs are rapidly attracting a stock of wild-fowl. Thus civilisation is in some measure restoring the balance of wild life, and offers to the most persecuted of our birds a quiet and secure retreat. I was able at the close of February, 1902, to witness a striking example of the results of wild-bird protection in increasing some species of wild-fowl which for half a century had steadily dwindled and disappeared, and were practically unknown anywhere in the neighbourhood of London. The scene was on the very large new reservoirs which lie between the grounds of the Ranelagh Club and the Thames, on what was some seven years ago a tract of market gardens and meadows. The construction of these lakes was so ably planned and carried out that in two years from the turning of the first sod four wide pools, covering in all one hundred acres of ground, were ready to be filled, and at the end of 1898 the ground was metamorphosed into the largest area of water in the London district, with the exception of the Serpentine.

It is so rare for changes of this magnitude to take place in any other way than by covering what was open ground with bricks and mortar, that the advent of a kind of reservoir flora and fauna so close to the greatest city of the world was looked for with some curiosity. All the waste ground not covered by the water or filtering-beds produced quantities of brilliant flowers, as waste ground enclosed and left to itself generally does. The banks and broad walks between the lakes were sown with good grass, which was regularly made into hay. The reservoirs themselves soon filled with fish, which came down the mains from Hampton, where the water is taken in from the river. What these reservoir fish found to live upon at first is not clear. No weeds are allowed to grow either in the water or on the banks, which are concreted. But the bottom becomes covered with the suspended matter deposited from the unfiltered water, and probably a considerable number of the minute *entomostraca* beloved of all fish breed in this. The Barnes reservoirs do contain a growth of weed, which is carefully removed every year.

Whatever their sustenance may be, these reservoirs are very full of fish, both the old ones at Barnes and the new lakes near Ranelagh. The supply of fish, and the open and strictly private extent of water, then attracted a number of wild duck or water birds of some kind, which the writer was invited to see and identify, as it did not seem probable that they could be the ordinary wild duck, which are vegetable feeders, and would need an artificial supply of grain, which is provided on the Serpentine, but is not given to any of these reservoir ducks. They have appeared entirely uninvited. The scene over the lakes was as sub-arctic and lacustrine as on any Finland pool, for the frost-fog hung over river and reservoirs, only just disclosing the long, flat lines of embankment, water, and ice; the barges floating down with the tide were powdered with frost and snow-flakes, and the only colour was the long, red smear across the ice of the western reservoir, beyond which the winter sun was setting into a bank of snow clouds. It was four o'clock, and nothing apparently was moving, either on the ice or the water, not even a gull. In the centre of the north-eastern reservoir was what was apparently an acre of heaped-up snow. On approaching nearer this acre of snow changed into a solid mass of gulls, all preparing to go to sleep. If there was one there were seven hundred, all packed together for warmth on the ice. It is on or about these reservoirs that the London gulls now sleep. Sometimes they are there in thousands; but the sealing of so much of the water with ice had sent a great proportion of them down the river to the more open water of the Essex marshes. Beyond the gulls, which rose and circled high above in the fog with infinite clamour, were a number of black objects, which soon resolved themselves into the forms of duck and other fowl. Rather more than seventy were counted, swimming on the water near the bank or sitting on the ice. These were the self-invited wild duck, so tame that with very little trouble they were approached near enough for their colour and form to be distinctly visible. The result of a look through the glasses was something of a surprise. They were not mallard, teal, or widgeon; but three-quarters of the number were tufted ducks, a diving-duck species, which haunts both estuaries and fresh water, but preferably the latter. It is a very handsome little black-and-white duck, seen in great numbers on certain large lakes in Nottinghamshire, and has greatly increased of late years in the county of Norfolk. But so far it has not appeared in any numbers either on the Surrey ponds or in Middlesex, and its assembling on this London reservoir is a remarkable proof of the tendency of wild-fowl to increase in this country.

The cock birds were in brilliant winter plumage, with large crests, white breasts, and white "clocks" on their wings. Some were sleeping, some diving, and others swimming quietly. When approached, the whole flock rose at once, and flew with arrow-like speed round the lakes and twice or thrice back over the heads of their visitors, of whom they were not at all shy, being used to the sight of the man who keeps the reservoirs' banks in order. They swept now overhead, now just above the ice, like a flock of sea-magpies or ice-duck playing before some North Atlantic gale. As several birds had not risen, we ventured still nearer, and saw that most of these were coots, some ten or eleven, which did not fly, but ran out on to the ice. Two large birds remaining, which had dived, then rose to the surface, and to our surprise and pleasure proved to be great crested grebes. These birds, which a few years ago were so scarce even in Norfolk that Mr. Stevenson despaired of the survival of the species as a native bird, have bred for three seasons in Richmond Park. But their presence so close to London shows that we need not despair of seeing wild-crested grebes appear on the Serpentine. These birds are so wedded to the water that they rarely fly. But this pair rose and flew, not away from, but towards us, passing within fifteen yards. With their long necks stretched out, feet level with the tail, and plumage apparently painted in broad, longitudinal stripes, they presented a very singular appearance.

The East of London owns a crowded wild-fowl sanctuary at Wanstead Park, which quite a different class of ducks frequent. It is now the property of the public, and very carefully administered by trustees. The lake there is very narrow and winding, which causes it to freeze easily. On the other hand, it is full of long, densely wooded islands, some almost enclosing pools of water. These islands shelter the birds, and when the lake is covered with ice the islands are crowded with wild duck and widgeon. Wanstead is a curious example of the faith of wild-fowl in a sanctuary, for the lake is so narrow that you could toss a stone among the fowl from the bank. Suburban houses are close by on all sides but the meadows by the little river Roding. Yet the fowl come to the lake as confidently as they do to great sanctuaries like Holkham. As there is a large heronry and rookery on the trees on the islands, the variety of life there is very great. The writer saw in weather like that in the second week of February, 1902, about a hundred and fifty wild duck, thirty or forty widgeon, a few teal, a pochard, and a great number of waterhens. Mallard, teal, dabchicks, and moorhens breed there regularly, and in hard weather a number of rarer birds drop in. Snipe are often seen by one of the shallower ponds, and occasionally such divers as goosanders appear and give an exhibition of fish-catching. These, like the tufted ducks and grebes, are entirely self-supporting. The wild duck are pensioners, being fed artificially, though they are wild birds, or descended from birds which were wild, just as are the London wood-pigeons.

THE CARRION CROW

Those familiar with the valley of the Thames and with the wild population both of the riverside and of the adjacent hills, will set down the carrion crow as the typical resident bird of the whole district. On the London Thames as high as Teddington it keeps mainly to the line of the river itself, on the banks of which and on the market gardens and meadows it finds abundant food, while the elms of large suburban residences give it both shelter and a safe nesting place. The bird is also commonly mistaken for a rook, and so shares the privileges of those popular birds. Higher up the river it swarms all along the Oxfordshire and Berkshire banks where not killed down by keepers, and a perfect army of them has for years invaded and been settled in the elm-bordered meadows of the Vale of White Horse. Thence it has spread on to the downs, where since the gradual abandonment of cultivation on the highest ground, and the removal of the scattered population of carters and keepers from a very large area, it now has matters all its own way. But it always haunted these heights, as the name "Crow Down," recurring more than once on the Ordnance maps, shows. The "Crow Down" with which the writer is less acquainted is on the very high, wild land north of Lambourn. There they have grown so confident that a nest was found in a thorn bush not ten feet high, at a place called Worm Hill, a good old Saxon name denoting that snakes abound there. There is no doubt that the crows kill and eat the young snakes, one having been seen carrying a snake in its beak recently.

The habits of the carrion crow are so independent and peculiar, and its resourcefulness so great, that it is not to be wondered at that it holds its own well within and around London, while the rook is gradually being edged out. It is generally regarded as a criminal bird, which it is to some extent in the spring. From that point of view the following facts may be cited against the crow. He is keenly on the look-out for all kinds of eggs about the time that his own nest is building. Consequently he is a real enemy to pheasants, wild ducks, plovers, moorhens, and other birds which lay in open places before there is cover. Nothing is more exasperating than these exploits to people who know where birds are nesting on their property, and wish to see them hatch safely. A wild duck's nest in a large copse was found by some persons picking primroses. In that copse was a crow's nest. The crows found out that the primrose-pickers had discovered something interesting, and a few hours later the "Quirk! quirk!" of the crows announced that they were enjoying life to an unusual degree. It was found that they had removed all seven eggs from the duck's nest. In an adjacent reclaimed harbour they took the eggs of ducks, plovers, redshanks, and even larks. In the Vale of White Horse they seem to take most of the early wild pheasant's eggs, besides stealing hen's eggs from round the farms. They are particularly fond of hunting down the sides of streams and canals in the early morning, where they find three dainties to which they are particularly partial,— moorhen's eggs, frogs, and freshwater mussels. They swallow the frogs *in situ*, and carry the moorhen's eggs and mussels off to some adjacent post to eat them comfortably. The shells of both eggs and mussels litter the ground under these dinner-tables. In Holland they are so mischievous that little "duck-houses" are made by the side of all the ornamental canals in private grounds for the ducks to nest in, a convenience of which they, being sensible birds, avail themselves. These duck-houses, or laying bowers, are still regularly made by the half-moon canal at Hampton Court, a survival probably of the days of William of Orange's Dutch gardeners.

[Illustration: THE LOBSTER SMACK INN, CANVEY ISLAND. *From a photograph by R. B. Lodge.*]

[Illustration: THE STEPPING STONES AT BENFLEET. *From a photograph by R. B. Lodge.*]

During the day they are very quiet birds, keeping much to the trees; but towards evening in March and April, their disagreeable croaking caw may be heard from all quarters where they are numerous. Just at dusk they become less wary than in the day. The writer for many years used to organise a few evening "drives" of the crows to try to thin them down before their ravenous families were hatched. Several guns used to hide in different parts of the valley near nests, and on to this "blockhouse line" the crows were driven. A few were generally shot before they discovered the plot. Solicitude for the nest seldom leads them into danger, but one pair met their fate in this way. The first bird came flying to the nest, in which there were eggs, as soon as a shot was heard in the distance. It was killed, and hardly had the spark of the flash disappeared when the other bird dropped down out of the gloom straight on to the eggs, and met the same fate. Forty young chickens were taken by a pair of crows from a farm in one spring. It was objected by some young ladies who were "interested" in the farm that the crows were "such sneaks." They used to come at luncheon-time up a line of trees extending from the wood to the farm. They were not in the least afraid of any one with a cart, apparently knowing that the horse could not be left, but would go straight for the chicken yard. A pair of sparrow-hawks near would seize a chicken now and then, but in a bold way as if they had a right to them. A few crows contrive to nest in Kensington Gardens. In the early mornings they always hunt the west banks of the Long Water, and are credited with taking a good many ducks' eggs, as well as ducklings.

Crows make one of the best nests constructed by the larger English birds. Usually it is placed, not out on the small branches, where rooks prefer to build them, but on the fork made by a large bough starting from the main trunk. This aids in concealment, and is a protection against shot, though probably the birds do not reckon on this contingency. The bottom of the nest is made of large, dead sticks. Upon and between these smaller twigs, often torn off green from willow-and elm-trees, or stolen from faggots of recent cutting, are laid and woven. Then a fine deep basin is made, woven of roots, grass, and some wiry stalk like esparto, the secret of where to find which seems a special possession of crows, and on this often a lining of bits of sheep's wool and cow's hair. There are sometimes as many as six eggs, and rarely less than four. They are quite beautiful objects, of a bright blue-green marked variously, but in a very decorative way, with blotches and smears of olive and blackish-brown. Two or three clutches of these eggs, with some of the splendid purple-red kestrels' eggs, and sparrow-hawks of bluish white, blotched with rich chestnut, make a very handsome show after a day's bird-nesting on the hills. The first eggs are laid very early, sometimes by the second week in April. A nest recently analysed consisted mainly of green ash taken from faggots and cuttings in the wood. One piece was a yard long, and as thick at the base as the little finger. The nest was *felted* with cow's hair, and quite impenetrable to shot. These nests last for years, and often have a series of tenants, kestrels, squirrels, brown owls, or hobbies. If the first nest is destroyed, the crow makes another. In his conjugal relations the carrion crow is a model bird. He pairs for life, and is inseparable from his mate. If one croaks, the other answers instantly, but usually they keep within sight of one another all day. In the evening the pair, seldom more than a few yards apart, may be seen hunting diligently in the meadows for slugs, which, so long as the weather is not too dry, form the regular supper of the birds.

A remarkable instance of the crow's courage in defence of its mate occurred some years ago on Salisbury Plain when a party were out rook-hawking. A falcon was flown at one of a pair of crows on favourable, open ground. The two birds mounted in the usual spiral until the falcon stooped, bound to the crow, and the pair came to the ground together. Just as the horseman rode in to take up the hawk the other crow descended straight upon the falcon, knocked her off its prostrate mate, and the two flew off together to cover before the falcon had realised whence the onset came. This crow not only showed great courage in facing both the falcon and the sportsman, but timed its interference with the greatest judgment and precision.

Probably a tame crow would make an amusing pet. Its intelligence must be very considerable, though the shape of its head does not so clearly indicate brain as does that of a raven. Among the crows which haunt the banks of the London river there are some highly educated pairs. One has maintained itself on the reach opposite Ham House for thirteen years, if the evidence used to identify them is reliable. These birds were noticed at that distance of time ago to have learnt to pick up food floating on the water. To see a big black crow hovering like a gull, and picking up bread from the bosom of the Thames, is so unusual that it always excites remark, and the writer was informed only last summer that these Ham House crows were seen doing this constantly. Not many years ago a crow nested in a plane-tree in St. Paul's Churchyard, and a pair also reside on the island in Battersea Park. But the great headquarters of London crows are the grounds of Ranelagh, and the reservoirs and market gardens of Barnes and Chiswick. They flock to the manure heaps in the latter, where the gulls now join them, and several pairs spend all day nearly all the year round on the reservoir banks at Barnes, and on Chiswick Eyot. The Eyot crows seems to find a good living there, and never leave it till their young, which are annually hatched in a tree at some distance on the Middlesex side, can fly. But the crows haunting the great Barnes reservoirs, where the tufted ducks now assemble in winter, are a bad lot. Last winter they were seen to single out and attack any gull separated from the flock which usually came there to roost. A sick or wounded gull was soon caught, killed, and eaten, the small black-headed gulls being no match for the crows. It was characteristic of their cunning that by the river itself they did not molest the gulls.

[Illustration: HAULING THE NETS FOR WHITEBAIT. *From photographs by R.
B. Lodge.*]

LONDON'S BURIED ELEPHANTS

The amount of river gravel left in the part of the Thames Valley on which West London is built is extraordinary. It is all round, and mostly red, and as there are no rocks like the stone which makes up most of this gravel anywhere in the modern valley, it is puzzling to know where it came from. I went to see the digging of the foundations of the new South Kensington Museum, and the great excavation, which was like the ditch of a fortress, and the stuff thrown out, which was like the rampart, was all dug in, or made of, river gravel. In this the men had found, lying higgledy-piggledy, with no two bones "belonging," quantities of bones of the beasts which used to graze on what I suppose was the Kensington "veldt," or perhaps flats by the riverside, during the time when the river's drift and brick earth was being deposited. The Clerk of the Works was much interested in these discoveries, and had caused them to be carefully collected. These were bones of the great stags then common, of the elephant, and of the primaeval horse, creatures which lived here before the Channel was cut between England and France, though not, perhaps, before man had appeared in what is now the Thames Valley, for flint implements are often found with the bones. Dr. Woodward, to whom some of the remains were taken, said that they reminded him of the great discovery of similar remains in the brick earth at Ilford, in Essex, thirty-seven years ago, when he personally saw, dug from the brickfields of that almost suburban parish, the head and tusks of one of the largest mammoth elephants in the world. These river-gravel and brick-earth buried bones are rather earlier than those found in the peat and marl. The latter belonged to creatures which, though they no longer exist in England, are still found in temperate Europe—beavers, bears, bison, and wolves. But the Thames gravel and the London clay are in places full of the bones of another, and earlier, though by no means primaeval, generation of mammals, some of which are extinct, while others are found at great distances from this country, in remote parts of the earth. Judging from the places where they are found and from the position of the bones, large animals must have swarmed all over what is now London, just as they do on the Athi plains and near the rivers and forests through which the Uganda Railway runs.

There was the same astonishing mixture of species, a mixture which puzzles inquirers rather more than it need. Hippopotamus bones are found in great numbers, and with the hippopotamus remains those of creatures like the reindeer and the musk ox, now found only on the Arctic fringe and frozen rim of the North, which lived on the same area and with them the Arctic fox. Judging from the great range of climate which most northern animals can endure, there is no reason to think this juxtaposition of a creature only found in warm rivers and of what are now Arctic animals is very strange. The London "hippo" was just the same, to judge from his bones, as that of the Nile or Congo. But the reindeer of North America, under the name of the woodland cariboo, comes down far south, and in the Arctic summer that of Europe endures a very high temperature. The Arctic fox does the same. If there were Arctic animals in Kensington and Westminster, that is no evidence that they lived in an Arctic climate. Looking over the list of bones, skulls, teeth, and tusks found, it is interesting to try to reconstruct mentally the fauna of greater London just previous to the coming of man. There were, to begin with, some African animals, either the same as are found on the Central African plains, and were found on the veldt of South Africa, or of the same families. The present condition of the country between Mount Kilimanjaro and the Victoria Nyanza shows quite as great a mixture of species. There, for instance, are all the big antelopes, rhinoceroses, zebras, lions, elephants, hyaenas, and wild dogs, and though there are glaciers on Kilimanjaro and the great mountains near the central valleys, the river running out of the Great Rift Valley is full of crocodiles and hippopotami. There is heather and, higher up, also ice and snow on the mountains, from whose tops the waters come that feed these crocodile-haunted streams. So on the London "veldt" there were lions, wild horses (perhaps striped like zebras), three kinds of rhinoceroses — two of which were just like the common black rhinoceros of Africa, though one had a woolly coat — elephants, hyaenas, hippopotami, and that most typical African animal the Cape wild dog! All these, except the elephants and hippos, can stand some degree of cold; and there is not the slightest reason why the two last may not have flourished in some deep river valley, very many degrees hotter than the hills above. To take an instance still remaining nearer to Europe than the Great Rift Valley. The Jordan Valley is very deep and very hot. Many species of birds are there found which are resident in India, and not anywhere nearer. It is a kind of hot slice of India embedded in the Palestine hills. The very large deer and immense bison and wild oxen probably fed on the same low veldt as the African animals. The bison were the same as those found in Lithuania, but far larger. Numbers of the skulls, of quite gigantic size, have been found in the brick earth. In the British Museum there is a tooth of the mammoth found in 1731, at a depth of 28 feet below the surface, in digging a sewer in Pall Mall. This Pall Mall mammoth might well figure in Mr. E. T. Reed's prehistoric series in *Punch*. Another tooth was found in Gray's Inn Lane. The mammoth was evidently not confined to the present region of clubland.

Besides these European and African groups of animals, a third class ranged the London plains, probably at a greater height and in a still colder temperature than the large grass-eating mammals mentioned. These creatures, whose bones are found plentifully in the drift, are now living in a country even more specialised than the African veldt. They are the creatures of the Tartar steppes and the cold plains of Central Asia. Their names are the suslik (a Central Asian prairie dog), the pika, a little steppe hare, and an extremely odd antelope, now found in Thibet. This is a singularly ugly beast with a high Roman nose, and wool almost as thick as that of a sheep when the winter coat is on. It must have been quite common in those parts, for I have had the cores of two of their horns brought to me during the last few years.

These dry bones are not made so astonishingly interesting by their setting in the gravel as are some far more ancient remains in England. The gravel is a mere rubbish-bed, like a sea-beach, in which all things have lost their connection. I was recently shown a set of fossils far more ancient, possibly not less than 2,000,000 years old, which were all found and may be seen exactly as they lay and lived when they were on the bottom of a prehistoric river which flowed through Hampshire, across what is now the Channel, over South France, and then fell into the Mediterranean. This river crosses the Channel at Hordwell cliffs on the Solent. There is the whole section, of a great stream two miles wide, with the gravel at its edges, the sediment and sand a little lower down the sides, and the mud at the bottom. On each lie its appropriate shells. Some are like those in the Thames to-day, but many more like those of a river in Borneo. They are so thick that out of a single ounce of the mud 150 little shells were obtained. In this, too, were found the tooth of a crocodile and the bones of a spiny pike, and in other masses of clay the very reeds and bits of the trees that grew there. These sedges of the primitive ages were quite charming. Even some of their colour was preserved, and all their delicate fluting and fibre, in the fine clay. One of the branches of a tree, now turned to lignite, had possessed a thick pith. This pith had decayed, and water had trickled down the hollow like a pipe. The water was full of iron pyrites, and had first lined the tube with iron crystals and then filled up the whole hollow with a frosted network of the same. There is a striking contrast between the presence and realism of these once living things still preserving the outer forms of life and the vast and inconceivable distances of "geological time."

SWANS, BLACK AND WHITE

A few pairs of black swans have been placed upon the river. Some of these rear broods of young ones, and appear to be quite acclimatised. The black swan was known to the traders of our own East India Company nearly a century before Captain Cook and Sir Joseph Banks discovered Botany Bay. The first notice of it appears in a letter, written about the year 1698, by a Mr. Watson to Dr. M. Lister, in which he says, "Here is returned a ship which by our East India Company was sent to the South Land, called Hollandia Nova," and adds that black swans, parrots, and many sea-cows were found there. In 1726, two were brought alive to Batavia, which were caught on the West Coast of Australia, near Hartop Bay, but no good account of their habits was ever written till Gould put together the facts he had seen and learnt on the spot.

The habits in their native land of birds which we only see acclimatised and domesticated, sometimes give a clue to what can be done to domesticate other breeds. This swan is only found in Australia, and only locally there, in the south and west. There it takes the place occupied by the Brent goose in our northern latitude, both as a water bird and as a source of food to the natives. "Wherever there are rivers, estuaries of the sea, lagoons, and pools of water of any extent the bird is generally distributed," says Gould. "Sometimes it occurs in such numbers that flocks of many hundreds can be seen together, particularly on those arms of the sea which, after passing the beachline of the coast, expand into great sheets of shallow water, on which the birds are seldom disturbed either by the force of boisterous winds or the intrusion of the natives. In the white man, however, the black swan finds an enemy so deadly, that in many parts where it was formerly quite numerous it has been almost, if not entirely, extirpated.

"This has been particularly the case on some of the larger rivers of Tasmania, but on the salt lagoons and inlets of D'Entrecasteaux's channel, the little-frequented bays of the southern and western shores of that island and the entrance to Melbourne Harbour at Port Phillip, it is still numerous." This was written in 1865, when to voyagers to the new continent the black swans of Melbourne Harbour were sometimes a first and striking reminder that they had reached a new world. One of the most deadly means of killing off the black swans was to chase them in boats, and either to net or club them, when they had shed all their flight feathers. This is what Mr. Trevor Battye saw the Samoyeds doing to the Brent geese on Kolguev Island. Thousands were driven into a kind of kraal, and killed for winter food. Next to the pelagic sealer, the whalers and ordinary seal-hunters are the worst scourges of the animal world. They killed off, for instance, every single one of the Antarctic right whales, and nearly all the Cape and Antarctic fur seals. But it is not generally known that they succeeded in almost killing off the black swans in some districts. They caught and killed them in boatloads, not for the flesh, but to take the swans' down. Black swans have white wings, though as they are nearly always pinioned here, a stupid habit which our people have learnt from the ancient and time-honoured brutality of "swan upping," we never see them flying. They are then very beautiful objects, with their plumage of ebon and ivory.

In Australia they begin to lay in October, and the young are hatched and growing in January. They are very prolific birds, laying from five to eight light-green eggs with brownish buff markings. Some years ago a splendid brood of six jolly little nigger cygnets were hatched out by the black swans at Kew. But the most successful breeder of black swans in this country was Mr. Samuel Gurney, who began his stock with a pair on the river Wandle, at Carshalton. He bought them in Leadenhall Market, in 1851. They did not breed till three years later, and laid their first egg on January 1st.

This is very interesting, because it shows that so far these birds were not acclimatised, but kept more or less to the seasons of reproduction proper to their native land. They were laying in what is the Australian summer and our mid-winter. It was a most severe winter, and the young ones were hatched out in a severe frost, which had lasted all the time that the birds were sitting in the open. The cygnets lived—it is not stated how many there were—and later on, the parents continued to breed, till in 1862, eight years after, they had hatched ninety-three young ones, and reared about half the number. The most extraordinary thing about the original pair was that they seem to have taken on both our seasons and their own, laying both in our spring and in the Australian spring, and so hatching two broods a year. They bred sixteen times in seven years—or probably seven and a half—and in that time laid one hundred and eleven eggs. The interest of this story is very considerable, because it shows the imperfect and exhausting efforts which Nature causes animals to make to adapt their breeding time to a new climate. Black swans which are descended from young birds bred in this country conform to the ordinary nesting-time of our hemisphere.

[Illustration: FISHING BOATS AT LEIGH. *From photographs by R. B. Lodge.*]

I notice that among the white swans on the Thames the cock-bird will fight to preserve his lady from intrusion, but he never thinks of taking her any breakfast, or of bringing her food of any kind, even though he may be fed most liberally himself. His only idea of helping her actively is by minding house while she goes off to feed and also while she is making her toilet. Not long ago, a swan who had a nest by the Thames so far forgot his mate as to fall in love with a young lady, whom he constantly tried to persuade to come and join him on the river. She was in the habit of feeding both swans every day, but as the lady swan was on the nest for the greater part of the time, the cock swan came in for most of the attention. In time he became tame enough to feed from her hand, and would come out on to the bank; but he preferred to sit on the water and to be fed from a boat-raft. After being fed he wanted to see more of his friend, but could not understand why she preferred stopping on such an uncomfortable place as the land when all she need do to enjoy his society, and to be happier herself, was to step down into the water. He would swim away slowly, looking over his shoulder to see if she was coming. As she usually wore a white dress, there is very little doubt that the swan thought she only wanted a few feathers to be quite a presentable swan, and suited for life on the river. When he found that she did not follow, he would return, and stretching out his neck would take hold of her dress and pull her towards the water, not in anger, but with a kind and pressing insistence, as showing her what was best. This he did usually when he had finished the food she brought, and when she left the bank would swim up and down, waiting to see if she were coming back.

The time-honoured brutality of swan-upping is now mitigated by law, its cruelty being obvious. It would be far better to leave them the use of their wings, which would enable them to seek food at a distance in winter, and to escape the ice, which sometimes breaks their legs. Several of these flightless swans were starved to death in 1902.

CANVEY ISLAND

Down near Thames mouth is the curious reclamation from the river mud known as Canvey Island. It is separated from the land by a "fleet," in which the Danes are recorded to have laid up their ships in the early period of their invasions, and the village opposite on the mainland is called Benfleet. Though on the river, it is a half-marine place, with the typical sea-plants growing on the saltings by the shore. In summer I noticed that the graves below the grey sea-eaten, storm-furrowed walls of the church have wreaths of sea-lavender laid upon them. But there is not the same rich carpet of sea-flowers as at Wells or Blakeney. Nor is the deposit so rich, so soft, so ready to be covered with smiling meadows as those of North Norfolk, built up from the mud-clouds of the Fen. Canvey Island itself is a heavy, indurated soil in parts, now well established, and producing fine crops. But is it the kind of ground which would pay a fair return on the cost of "inning it" to-day? The wheat is good, the straw long, and the ears full. The oats are less good, perhaps because the soil is too heavy. The beans are strong and healthy; clover, which does not mind a salty soil, thrives there; and there are strong crops of mangold. But it is not like the Fenland; it cracks under the sun, "pans" upon the surface, and is not adapted for inexpensive or for intensive cultivation. Such was the writer's impression from a careful view of the farms in the middle of harvest. But as a fact in the history of English agriculture, and in its relation to the past story of the Thames mouth, and its possibilities as a future health resort, this work of the enterprising Dutchmen in the beginning of the seventeenth century is full of interest. In 1622 Sir Henry Appleton, the owner of the marsh, agreed to give one-third of it to Joas Croppenburg, a Dutchman skilled in the making of dikes, if he "inned" the marsh. This the Dutchman did off hand, and enclosed six thousand acres by a wall twenty miles round. Like many parts of the Fens, the island was peopled for a time by Dutchmen engaged on the works, and Croppenburg is said to have built there a church. Two small Dutch cottages remain, built in 1621. The general aspect of the island is like that part of Holland near the mouth of the "old" Rhine, but less closely cultivated and cared for.

[Illustration: THE LOBSTER SMACK INN, CANVEY ISLAND.
From a photograph by
R. B. Lodge.]

[Illustration: THE STEPPING STONES AT BENFLEET. *From a photograph by R. B. Lodge.*]

It has always been a separate region. Never yet has it entered the heads of its proprietors to join it permanently to the mainland. For three centuries its visitors and people have driven or walked over a tide-washed causeway at low water, or ferried over at high tide. You do so still, in a scrubbed and salty boat, while an ancient roadmender is occupied in the oddest of all forms of road maintenance. He stirs and swirls the mud as the tide goes down, to wash it out of the hollow way, otherwise it would be turned into warp-land every day, and become impassable. The Dutchmen's roads are sound and straight enough on the island. Outside the wall the samphire and orach beds are wholly marine. Inside the dikes and ditches are filled with a purely sweet-water vegetation. Further seawards, or rather riverwards, at a place called "Sluis," they are fringed with wild rose and wild plum, and the ditches are deep in rushes, in willow herb, in purple nightshade, water-mint, and reeds.

Camden gives a curious account of the island in his day. It was constantly almost submerged. The people lived by keeping sheep on it. There were four thousand of a very excellent flavour. Evidently this was the origin of *pré-salé* mutton in England. Camden saw them milking their sheep, from which they made ewe-milk cheeses. When the floods rose the sheep used to be driven on to low mounds which studded the central parts of the marsh, and these mounds are there still. Some are covered with wild-plum bushes. One, in the centre of the island, is the site of the village of Canvey; and on one, at the time of the writer's last visit, two fine old Essex rams were sleeping in the sun. There was no flood; the island had not known even a partial one for some years. But true to the instincts of their race, they had occupied the highest ground, though it was only a few feet above the levels. There are few land-birds on Canvey Island, because there are few trees. Some greenfinches, a whinchat or two, almost no pipits or larks, and very few sparrows. The shore-birds are numerous and increasing, for the Essex County Council strictly protect all the eggs and birds during the breeding season. Enormous areas of breeding ground are now protected in the wide fringe of private fresh-water marshes of this river-intersected shore. Plovers, redshanks, terns, ducks, especially the wild mallards, are increasing. So are the black-headed gulls; even the oyster-catchers are returning. After nesting the birds lead their young to the southern point of Canvey Island. It is too near the growing and popular Southend for the birds to be other than shy. But as they are not allowed to be shot till the middle of August, they are able to take care of themselves. At the flow of the tide, before the shooting begins, the visitor who makes his way to this distant and unpeopled promontory sees the birds in thousands. Out at sea the ducks were this year as numerous as in the old days before breechloaders and railways. Stints and ringed plover, golden plover and redshanks were flitting everywhere from island to island on the mud and ooze; curlews were floating and flapping over the "fleets"; and all were in security. As the tide flowed, they crowded on to the highest and last-covered islets, whence, as the inexorable tide again rose, they took wing and flew swiftly to the Essex shore. The Sluis, looking across to the Kentish shore, is the home of the seagulls. Many quaint ships lie anchored there—Dutch eel-boats, which call for refreshment after selling the cargo; barges; hoys from the Medway bound to Harwich; and fishing-smacks and timber-brigs. Round these the seagulls float, as tame almost as London pigeons. They prefer company, at least the lesser gulls do; the big herring gulls and black-backed gulls keep aloof.

[Illustration: HAULING THE NETS FOR WHITEBAIT. *From photographs by R. B. Lodge.*]

The hope of reclaiming land from the waves exercises a peculiar fascination over most minds. It presents itself in more than one form as a most desirable activity. It is something like creation—a form of making earth from sea. The clothing of the fringe of ocean's bed with herbage, the reaping of a harvest where rolled the tide, the barring out of the dominant sea, the vision, not altogether illusive, of planting industrious and deserving men on the ground so won, all these are alluring ideas. The undertaker, to use the word in vogue in the Stuart days when such enterprises were in high favour, always leaves a name among posterity, generally an honoured name, and in nearly every case one associated with courage, perseverance, and in some measure with benevolence. The picturesque and sentimental side will always remain to the credit of the reclaimers of the waste of Neptune's manor. But if the balance of profitable expenditure, or of good done to others, is weighed between winning land from the sea and expenditure in improving the cultivation of land already accessible, the award should probably be given to the latter. Intensive cultivation and the improvement of the millions of acres which we now possess is a more thankworthy task, demands more brains, and should give greater results than the gaining of a few thousands of acres now covered by water. This conclusion is not the one which any lover of enterprise or of picturesque endeavour would prefer. It is a pity that it is so. Perhaps in days to come when wheat is once more precious the sea wastes may once more be worth recovery. But even so they are not desirable spots on which to plant a population. They are by natural causes on the way to nowhere, and out of communication with the towns and villages. Brading Harbour, in the Isle of Wight, is an exception, for it ran up inland. Lord Leicester's marshes at Holkham are narrow though long, and, while splendidly fertile, are all well within reach of the farms and villages. But to scatter farms and labourers' cottages on the dreary flats of a place like Canvey Island is not likely to appeal to the wishes of modern agriculturists, who feel the dulness of rural life acutely already. The growth of the Jewish colonies not far off on the mainland, where poor Hebrews continually reinforce a community devoted to field and garden labour and content to begin by earning the barest living, seems to indicate that a population from the poorest urban class might be found for reclaimed land. But the industrious town artisans of English blood have not yet found life so intolerable as to be ready to try the experiment.

THE LONDON THAMES AS A WATERWAY

Mary Boyle, in "Her Book," speaking of the time when her father had an appointment at the Navy Board and a residence in Somerset House, says, "It was our great delight to go by water on Sunday afternoon to Westminster Abbey, and there is no doubt we occasionally cut a grand figure on the river; for when my father went out he had a splendid barge, rowed by boatmen clad entirely in scarlet, with black jockey caps, such as in those picturesque old days formed part of that beautiful river procession in honour of the Lord Mayor, on the 9th of November, over the disappearance of which pageant I have often mourned."

It was not until the early days of the present reign that neglect and dirt spoiled our river as an almost Royal waterway; and we believe that as late as the days of Archbishop Tait the Primate's State barge used to convey him from Lambeth Palace to the House of Lords opposite. State barges and river processions were the standing examples of State pageantry, thoroughly popular and remembered by the intensely conservative people of London; and it is a tribute to the feeling that the use of the river was a necessary part of London life, that the Lord Mayor and his suite on the 9th of November used to take boat at Blackfriars Bridge, and went thence by water to Westminster Hall, returning in their State barges to the bridge, where their coaches were waiting for them. We may credit the founders of the earliest illustrated paper with a knowledge of the popular sentiment of the day. When the *Illustrated London News* was established the title-page of that paper showed the Thames, with the procession of State barges in the foreground, and the then new and popular river steamers passing by them.

In addition to cleanliness something in the form of a restoration of old conditions of water-level and other improvements by modern engineering will also be required if the river is to become a popular waterway. Among the main drawbacks to its present use is the great difference in level between high and low water. The old London Bridge, with its multiplied arches and pillars, acted as a lock. It admitted the flood tide more easily than it released the ebb. The consequence was that when the tide began to fall the waters above were pent in by the bridge, and the river was kept at a level of three feet higher than it was below the obstruction. Even now at flood tide it is a splendid and imposing river. But the very improvements which add to its dignity when the tide is flowing, have caused it to remain almost waterless for a longer period during each day. The dredging and deepening of the channel forces the waterway to contract its flow, while the embanking of its sides enables the tide to slip down at great speed. For four hours in each tide the Thames is not so much a river as a half-empty conduit. It is not in the least probable that this will be allowed to continue. The success of the half-tide lock at Richmond has been beyond all expectation. It has secured a perpetual river, whether on the ebb or flow, with a mean level suited for boating and traffic at all hours. A scheme for another lock of the same kind at Wandsworth is now accepted in principle and nearly completed in detail. When this is built the long stretch of river from Wandsworth, past Putney, Ranelagh, Hammersmith, Barnes, and Kew, will retain a permanent and constant supply, augmented at the flood tide, but never falling below a certain level at the ebb. Then must follow the final and complete measure for making the London river the greatest natural amenity in the Metropolis, a half-tide lock at London Bridge, to hold up the water opposite the historic and magnificent frontage of St. Paul's, the Temple, Westminster and Lambeth, and upwards to above the embankments at Chelsea. The result would be an immense fresh-water lake, with an ebb and flow to keep it sweet and pure, but remaining for the greater part of the twenty-four hours at a fixed level, and during this period of rest only moved by a very gentle downward stream, or else practically still when the water sank level with the sills of the lock. This would make it not only easy for boats propelled by steam, sail, or oars to move on it at all hours, without hindrance from the present strong up or down currents, but also absolutely safe. Any craft, from the outrigger and Canada canoe, to the improved river steamers which would at once be launched upon its waters, could float with ease and safety on the London Thames.

The scene in the near future can be imagined from the analogy of Henley, though the larger scale of the London river makes the forecast more difficult to bring into proportion. The intentionally decorative side, given on the upper river by the houseboats, will doubtless be supplied by a new service of public or municipal passenger steamers, able to ply continuously at all hours, independently of the tide, as fast as safety permits, and absolutely punctual because the stream will be under control. These should be as brilliantly carved, gilded, coloured, and furnished as possible, surplus profits only going to the municipal coffers after the boats have been repaired yearly and thoroughly redecorated. The scheme is not in the least visionary. The Chairman of one of the tramway companies obtained recently complete estimates for a fast, luxurious, and beautiful service of Thames passenger boats, which he was convinced would pay even now; and though he did not succeed in inducing the shareholders to accept the idea of this alternative investment, there is no doubt that on the improved river the improved steamers would pay. A simultaneous and necessary addition would be the building of numerous broad, accessible, and beautiful stairs and landing places. Instead of the narrow gangway through which files of passengers slowly creep there must be long platforms, on to which the crowds on board the vessels step, as from a train, all along the length of the ships, so that the touch and departure may be rapid. The decline of traffic on the river is largely due to the narrowness and fewness of these points of access, which were gradually closed as the river was deserted for the road, while their blocking or neglect discouraged efforts to improve or multiply boats and steamers.

In 1543 there were twelve large and handsome flights of stairs down to the river between Blackfriars and Westminster. In 1600, besides these there were public and private gateways of large size, covered docks for State and private barges, and every convenience for access to the water. There were stairs and stages at Essex House, Arundel House, Somerset House, York House (the water-gate of which still remains, with a frontage of embankment and garden between it and the river of to-day), Bedford House, Durham House, Whitehall, and Westminster. The latter were "the King's Stairs." There are few constructions which lend themselves better to architectural treatment than water-gates and stairways. They would become one of the features of the Embankment. On the river itself the City Companies would once more launch their State barges, and the Houses of Parliament would have a flotilla of decorative steam or electric launches. Permanent moorings, now difficult to maintain near the bank on account of the runaway tide, would hold boats, launches, and single-handed sailing yachts. No one will grudge the County Council a State barge; while the new municipalities which border on the river—Westminster, Southwark, Fulham, Kensington, and the rest—will endeavour to interest their members in the great waterway by following the example of the Thames Conservancy and sending their representatives for official voyages to survey its banks and note suggestions for improvements in their actual setting and surroundings. No doubt in winter all the minor pleasure traffic would cease. But there is no reason whatever why a service of ornamental and well-equipped screw steamers plying at very short intervals, and with absolute punctuality, should not continue all the winter through. They would be entirely unlike the "penny boat." Double-storied deckhouses, glazed and warmed, would afford the passengers more room, purer air, and a more rapid means of transport than the omnibus, and a far more agreeable mode of crossing from one side of the river to the other than by railway bridges, tunnels, or the architecturally beautiful, but crowded, stone bridges used for ordinary traffic.

THE THAMES AS A NATIONAL TRUST

A movement is on foot among various societies interested in the preservation of outdoor England to take measures jointly for the protection of the beauties of the Thames. The subject is one which attracts more interest yearly, and the time has now come when the nation should make up its mind on the subject of such splendid properties as it possesses in "real estate" like the Thames and the New Forest, with especial regard to their value for beauty and enjoyment. It would be unfair to expect too much from the Thames Conservancy in this direction. That body exists to maintain the navigation of the river, and to see that no impediments are put in the way of its use as a waterway. Its duties are, in the first instance, those of a Highway Board, which deals with a river instead of a road. It has to buoy wrecks, and see that they are raised. It controls the speed of steamers and launches, not, in the first place, because they are a nuisance to pleasure boats, but because the "wash" destroys the banks, and this costs money to repair. It arranges for the dredging of shallows in the fairway, for the embankment of the shores, and for the repair and maintenance of the locks. Its business is to do this as cheaply as is consistent with efficiency, and to lay no unavoidable burden on the trade of the river. The preservation of its amenities is not, strictly speaking, the object for which the Conservancy exists. Yet it has done much in this direction, by obtaining from time to time powers not originally in its jurisdiction. It may be said to be on its way to become a guardian of the amenities of the river, though these, which are fast becoming far more important than its use as a means of traffic, were at first only accidentally objects of solicitude to the Conservators, and such attention as is by them devoted to this end is mainly confined to the Upper Thames, and not to the London river. Legislation to preserve natural beauty, or prevent disfigurement, has practically only been possible in recent years, and the wish to do so, though shared by most classes, is not yet so pronounced as it ought to be. What the Conservancy has been able to do, under these circumstances, has been done, partly on grounds of health, which are recognised in Legislation, and partly to preserve the fishery. It has endeavoured to keep the river from the most disgusting forms of pollution, and lately from being made the receptacle for minor but objectionable refuse. It has certainly prevented the Upper Thames being made into a sewer, and also stopped pollution by paper mills and factories. London's need of pure drinking water has given immense assistance to the forces which were working to keep our rivers clean. All the tributaries of the Thames are now under surveillance, and no village or little country town may use them to pour sewage into. Country villagers may grumble at being forced to keep water clean for Londoners to drink. But this Act has done more to preserve the amenities of the countryside than any other of this generation. It is so far-reaching, and so frankly expresses the principle of placing public rights in the "natural commodity" of pure water in our rivers before private convenience in saving expense, that it is a hopeful sign of the times.

While the existence of this extensive control is a guarantee for the increasing pureness of the Upper Thames, it is also a precedent for regulating and increasing the supervision of this national property in the most beautiful, the largest, and the most pleasant highway in our country, whose very pavement is a means of delight to the eye, of pleasure to the touch, and of refreshment to all the senses. The minor regulations for its maintenance are still more encouraging, for some of these aim directly at preserving beauty, or objects of natural interest, for their own sake. The oldest are those which protect the fishery. There is one close-time for the coarse fish, another for the trout, and a limit of size to the meshes of the nets which may be used. Such minor disfigurements as the throwing of ashes from steam-launches into the water or of kitchen *débris* from houseboats are forbidden. Recently the Conservators have taken powers more frankly directed to the preservation of natural beauty, though even in these cases what may be called direct "taste legislation" has not been exercised. They have not asked for leave to say definitely: "This or that object is hideous or disfiguring, and cannot be allowed by the side of our national highway." But they have said, "This or that object which grows on or lives by the side of our river-road is beautiful, and gives pleasure to the public, and therefore it shall not be destroyed." The result has been that the birds on the river and its banks may no longer be shot, and certain flowers are not permitted to be plucked. The Conservancy is also able indirectly to exercise some control over riverside building operations, and very recently compelled an alteration of design in the use of a building site on a reach of the Upper Thames.

[Illustration: FISHING BOATS AT LEIGH. *From photographs by R. B. Lodge.*]

It may be asked why, if so much has already been done, we should not rest contented with the present control of the river, trusting that a gradual increase of powers will be granted to the Conservancy, so that little by little they may be able to meet all requirements for the preservation of the Thames as our national river, just as the New Forest is preserved on the grounds that it was "of unique beauty and historic interest."

The answer is that, in the first place, this is not the proper business of the Conservancy, but only an incidental duty; and, in the next, that with the best of goodwill, as is shown by what they have done, the Conservators have only been able to mitigate, not to control, a vast amount of disfigurement and abuse of the river in the past. They were not created *ad hoc*, and the body has not the position which would enable them to take a strong line, or powers for expenditure on purely non-remunerative business, such as might be necessary if a millowner had to be bought out if about to sell his property for conversion into a gasworks, like the factory of the Brentford Gas Company just opposite the palace at Kew, or the foul soapworks which for years disfigured the banks and polluted the air at Barnes. They have not the funds to maintain a proper police to stop the minor pollution of the river, or to scavenge it properly, and anywhere below Kew Bridge they are entirely unable to cope with bankside disfigurements. Else we cannot believe that for years the bank opposite the terrace at Barnes and the villas above it would have been given up to the shooting of dustbin refuse for hundreds of yards, or that Chiswick and Richmond would have been permitted to pour "sewage effluent" into what are still two of the finest reaches on the London river, or that we should see advertisements of "A Site on the River— Suitable for a Nuisance Trade," advertised, as was recently done, in a daily paper. If the London public, for instance, will only make up its mind in time that the Thames is something really necessary to its enjoyment of life; that it is the most beautiful natural area which they can easily reach; that on it may be had the freshest air, the best exercise, good sport (if the fishery were replenished and the water kept clean), and constant rest and refreshment for mind and body—it would no doubt succeed in inducing Parliament to put the river under a strong Commission with an adequate endowment. But the preservation of the Thames is more than a local, or even a London, question. It is a national property and of national importance, and should be managed from this point of view. Mr. Richardson Evans has made out a good case for national *property* in scenery generally. But here the case is stronger, because the river *is* a national property already, and anything which decreases its amenities for private ends damages the property. Like very much other real estate, its value depends now not on its return to the nation as a highway (above London, that is), but purely as a "pleasure estate." Supposing any private owner to be in possession of a beautiful stretch of river, is it conceivable that, if he could, he would not get a law passed to prevent gasworks, or hideous advertisements, or rowdy steamers, or stinking dust-heaps, or sewage works from spoiling any part of it? Would he let people throw in dead cats and dogs, or set up cocoa-nut shies on the banks? — all of which things have been done, and are done, between Syon House and Putney Bridge, on the way by river from London itself to London's fairest suburbs, Richmond and Twickenham. Or would he allow himself to be shut off from access to his own river, or forbidden to walk along the path by its side, supposing that one existed? Yet the public, whose rights of way on the Thames are as good as those of any private owner on his own waters, either suffer these things to go by default, or at most permit and only faintly encourage a body which was not created to care for this purpose, to undertake it because there is no other authority to do so. It is no use to leave these things to the local authority, however competent. There is always the danger that local authorities — even those representing interests normally opposed to each other — may agree to press local interests at the expense of the public. What is needed is that both the New Forest and the Thames shall be created national Trusts. Both are as valuable, as unique, and as important as the British Museum, and should be controlled by trustees of such standing and position that their decision on matters of taste and expediency in managing and maintaining the natural amenities of the national forest and the national stream would be beyond question. The decisions of the trustees of the British Museum are scarcely ever questioned by public opinion. Could not the national river be placed under similar guardianship?

www.ingramcontent.com/pod-product-compliance
Lightning Source LLC
Chambersburg PA
CBHW050215230526
45470CB00001B/397